In Quest
of Telescopes

In Quest
of Telescopes

By
MARTIN COHEN

FIRST EDITION

SKY PUBLISHING CORPORATION
Cambridge, Massachusetts

Library of Congress Cataloging in Publication Data

Cohen, Martin, 1948-
   In Quest of Telescopes.

   Includes index.
   1. Telescope. 2. Astronomy — Vocational guidance.
I. Title.
QB88.C65   522'.2'0924 [B]   80-14057
ISBN 0-933346-25-5

First printing, 1980

Printed in the United States of America

*To Elizabeth, the most unexpected and
welcome consequence of my quest.*

CONTENTS

ILLUSTRATIONS

PREFACE

All you have to do is stand outside on a dark clear night and look up at the sky. Aren't there questions that come into your mind when you look at the stars and notice their different colors, brightnesses, and patterns? In some of us, those questions keep growing and becoming more insistent until we have to get involved in astronomy, at least as amateurs.

This book is really my astronomical autobiography, as far as 1979. It tells you what it is like to lead the life of a professional astronomer. I know that during my late teens I would have liked to find and read such a book, but none was available. I proffer only one piece of additional information — if what drives me to do astronomy drives you too, then you will never regret the decision if you enter astronomy as a career. But keep a sharp eye on the job market. If there comes a time when you want stability, security, a guaranteed income, beware of the perils of astronomy. There are few permanent academic positions open each year, and the financial pies are being sliced ever more thinly. Have a second career somewhere in mind if your astronomical opportunities begin to dwindle.

It is hard for me to acknowledge here all those people who have made my astronomy the particularly enjoyable experience it has been during the last decade. There are astronomical friends whom I shall value throughout my life; colleagues who have sweated out cloudy nights with me; students who have helped me immeasurably and unselfishly on observing trips; people I see only once or twice a year with whom it is always a delight to meet and exchange news and ideas; friends who have stood by me in various crises. But I do want to thank those who have had the greatest influence on me: my parents, David and Carol Allen, Gordon Keenay, David Foster, Douglas Heggie, Peredur Williams, Patrick Moore, Nick Woolf, Fred Gillett, Tom Murdock, Bill Fawley, Michael Barlow, Dick Treffers, Leonard Kuhi, Mal and Marilyn Raff, Dick Schwartz, Michael Merrill, Nuria Calvet, Stuart Vogel, Bob Lewis, Frank Holden. I wish you all clear skies.

Martin Cohen,
Oakland, California, 1979

BEGINNINGS 1

I think it happened when I was about four years old. I had a vivid imagination and an excellent visual memory, and I was apt to remember all manner of trivia, like comic strips. It was a cross between *The Adventures of Richard Lion*, a curiously articulate feline who was forever sailing off into space, and *The Golden Book of Astronomy*, a lavish, full-color primer in astronomy, that eventually hooked me. I would spend hours attempting to visualize conditions on our Moon or on Venus, or imagining spaceflight. Slowly, insidiously, my flights of total fantasy were colored by an increasing proportion of facts. It was then a simple matter of raiding school and public libraries, devouring everything vaguely pertaining to astronomy and to space.

My first practical experimentation was launched because of the hoarding instincts of my grandmother who had secreted away a fine collection of ancient spectacles. In the early days of optics and optometry the toric lens had not been invented, so spectacle lenses were circles of glass, axially symmetric in their optical properties. In brief, they were admirable candidates for the objectives and eyepieces of small telescopes. I began with a simple Galilean system, having a magnification of about three, whose power I eventually boosted to ten. This elementary device was sufficient for my initial efforts in mapping the heavens and learning the patterns of the stars.

The second phase in my amateur astronomical career was wrought by my attendance at the Manchester Grammar School, which owned a fine, old, brass telescope made in the early 20th century. It was a refractor, with a 3½-inch achromatic objective, and it boasted a fascinating array of gleaming brass accoutrements. There was a Herschel wedge, for safe solar observation, that permitted ninety-six percent of the sun's light and heat to pass along the optical axis of the objective, while diverting a few percent to the eyepiece by reflection from an unsilvered piece of glass. There were eight fine old eyepieces, from a spectacular wide-field 20x through Ramsden designs to about 400x. The telescope was mounted on a monumental tripod with an enormous and extremely heavy brass altazimuth cradle, and it was equipped with retractable bracing rods that supported the tube in a wind by attaching it severally to the wooden tripod legs. It taught me the first law of practical astronomy — develop your arm muscles. It was also a wonderful instrument that must have made its constructor very proud of his work.

Many an enthralling evening did we members of the school astronomical group spend; mapping the surface of our Moon at 270x; collecting colorful close double stars; tracing the patterns of Jupiter's moons about the yellow, detailed globe; or just browsing at 20x through the rich starfields of Auriga, Orion, and Cygnus. By that stage I felt inextricably bound to astronomy, although clearly I did not perceive in what way I would enter that field. This feeling was reinforced when I obtained my own telescope, a small, 4-inch reflector, well-designed for lunar and planetary work, and for learning the gentle art of astronomical sketching. Actually, the telescope, purchased from a supposedly reputable optical shop in the city of Manchester, possessed a severe limitation and one that perplexed me literally for years. Its images were atrocious, full of apparent coma and other bizarre distortions that I could not categorize. Of course, I still spent hundreds of cold, English, winter nights in the back garden tinkering with the constellations. Eventually, once I had learned the rudiments of mirror-grinding and testing, I found that my primary mirror had two quite different spheres ground in it, each covering about half the total surface area. By stopping it down near the Newtonian flat I was ultimately able to clean up the images to a stunning degree. At that phase I was also concocting composite eyepieces of my own design by demolishing and cannibalizing all manner of simpler lenses.

In retrospect, those years spent with the 3½-inch and 4-inch telescopes were fundamental to my subsequent thinking. They instilled in me a profound intuitive respect for the beauty of the heavens, its nebulae, its rich, compact star clusters, the orderly nature of planetary motions, the detail of the lunar surface. Those years also set the basis for patient sketching of what my eye and brain told me I saw through the eyepiece. They told me that I wanted to be an astronomer. They gave rein to my youthful masochistic tendencies and enabled me to ignore moderate cold, wind, or discomfort just to acquire a sketch of, or perhaps merely a view of, another cluster, nebula, or lunar crater. I had a nodding acquaintance with other telescopes in my grammar school years, but none could be said to have impressed me by its character so much as the handsome 3½-inch and its rather more ramshackle cousin, the 4-inch.

Perhaps the next most significant telescope I met in that period was the school's 9-inch reflector. This instrument had had a checkered career, beginning life as an 8-inch blank (nautical porthole glass, I recall) that was ground in a school darkroom by us enthusiasts. The little astronomical society in school was run each year by some pundit in his last or penultimate year at the school. The variable status of this second school telescope was very much in the hands of this chairman. Two years preceding my reign in this exalted role, the chairman had been a preeminent observer who had constructed a mirror cell and a spider mount for the secondary. My immediate predecessor, on the other hand, was a youth more disposed to the grinding and testing of mirrors than to their completion and use, although precisely the person to grind, test, and polish a blank and convert it into a mirror. And so the partially figured 8-inch gave rise to the 9-inch. Consequently, by the time I became chairman, what was required was a tying together (metaphorically, I hasten to add) of the disparate components, and the construction of a suitable equatorial mounting.

After months of delicate negotiations with sundry sports committees, designed to ensure the construction of a brick and concrete plinth to which we could affix the head of this mounting, such a plinth appeared, almost miraculously, one evening, between the First Eleven cricket pitch and the adjoining Rugby field. We had laid a three foot square of concrete ourselves, but for the plinth we were not responsible. The school bricklayer was an elderly grizzled fellow whose job resembled that of a Golden Gate bridge painter. He would tend to the brick and mortar work of the extensive school buildings from foundation stone to bell tower, and then would recommence a new cycle on completion of the old. Apparently this bricklayer got wind of our general idea and simply took it into his head to lay a square section brick edifice on our concrete bed. After several minutes of just appreciation of his labors, I can recall a chill marching up my spine as I realized that we had yet to set a square frame with four protruding bolts onto the plinth, pointing celestially north. No one had even suggested to the bricklayer that this might be a constraint. A compass was rapidly procured and after some tense moments a government survey map of the area was located, in the dog-eared margin of which the geomagnetic deviation had been imprinted. That bricklayer must have been unusually sensitive to magnetic fields, for we were able to align our mount to celestial north after only a very slight rotation with respect to the plinth section.

That was a proud moment, one Friday evening, as I marched at the head of a column of boys of monotonically decreasing height. Resolutely we stalked between the cricket and rugby fields. At the front were boys of large girth and proportionate wisdom, staggering along with the massive steel equatorial head; behind them came middle-schoolers carrying the open-lattice angle-iron tube as pall bearers might; next came a lad sweating from prodigious concentration and bringing the mirror in its padded box. In the rear were a succession of first year pupils, bearing a host of diminishing items, from several barbell weights (filched from the gymnasium to do service as counterweights) to the eyepieces and assorted tools for tightening nuts in awkward corners, thoughtfully conceived of and manufactured by myself.

Yes, it was a proud, but also a tense, evening. In an hour we had erected the assorted carpentry and metalwork into a telescope; not the most rigid or elegantly streamlined of devices, but a functioning instrument. We saw the moon, as I recall, then it clouded over and we contented ourselves with resolution tests by viewing the lighted windows of a nearby university dormitory. We were undoubtedly the most scientific of voyeurs that night as we peered at chromatically tinged faces and successively realigned the optics.

It is in the nature of all academic institutions that their populations are transient. So it was that I, and my contemporary cohorts in astronomy, eventually lost contact with the school telescopes. However, I was on my way to professional astronomy although, again, I could not have predicted in precisely what manner. My first set of national examinations spanned a catholic array, from mathematics to

Latin and Greek (all amateur astronomers should have a head start in Greek, having acquired at least that ancient alphabet from star charts). My advanced examinations (pre-university level) were more specialized, mathematics and physics alone, and it was as a student of mathematics that I continued my academic career and simultaneously my amateur astronomy.

Almost eight months passed between the time I left high school and the time that my university life began. During that period I was out in the real world, as opposed to the academic, working essentially as a clerk. I was also working on one astronomical project in my evenings. My father had permitted me to use an old bellows camera he had picked up during the mists of World War II. I had become fascinated by the idea of sky photography as a means of producing star charts to limits fainter than Norton's *Star Atlas* (about sixth) or the *Atlas of the Heavens* (nearly eighth). But what I really wanted was an equatorial head, driven at sidereal rate, so as to concentrate all the starlight into a spot on the emulsion, rather than relying on the messier (no astronomical pun intended) star trails. During my final months at grammar school, I had constructed a number of curiously shaped aluminum oddments, intended to act as the core of this astrophotographic device. At the heart was a small geared motor whose output shaft revolved once in twenty-four hours.

Oh, it was a thing of beauty! The base was a heavy wooden and cardboard tube about four feet high from a roll of carpet. Inside the plinth I had small transformers that provided low voltage for small lamps so I could record what exposures I had taken, also for a fat resistor taped inside the camera lens hood to prevent dewing. On top of the base sat a stubby cast-iron shaft that was moonlighting during the daytime as the base of a heat lamp against my grandmother's arthritis. My own ingenious metalworkings lived on top of this shaft, were adjustable for latitude, azimuth, held the motor and a clutch assembly, and supported camera and counterweights. Exposures were dialed onto a mechanical egg-timer, but, on completion of the allotted time, instead of ringing a bell, the hammer was arranged to release a spring holding the plunger of a cable release connected to the camera shutter. I know it sounds Rube Goldberg; it was, but it all functioned! Of course, the star images were never diminutive elegant spots; they were shaped like crotchets, due to ill pointing at the north pole and to backlash in the drive train of the motor. Nevertheless, it did wonders for my engineering self-confidence.

2 CAMBRIDGE UNIVERSITY

A major step toward my career in astronomy came in October, 1966, when I "went up" to Cambridge University. The university astronomical society was traditionally an active one, with a total annual membership that exceeded 300 during my undergraduate career. Within this microcosm of the astronomical world it was easy to distinguish the aficionados, who were active observers, from the budding theoreticians. The latter attended lectures and socials, but their feet rarely inclined them along Madingley Road to the Observatories.

The society had use of several instruments housed on the site of the historically famous observatories, the largest among these being The Great Northumberland Refractor. This was the largest refractor in the world for a period in the 19th century, after which it was displaced in that rank by the Harvard 15-inch. The Northumberland, as it was affectionately termed, had been endowed in 1833 by the Duke of Northumberland who, needless to say, did not reside in the north. It is a fine, ancient, refractor of almost 12 inches aperture, whose present objective (since 1950) is the world's largest triple-component air-spaced achromat. Of unusual design, the long (f/20) tube is square-sectioned and made of mahogany (believed at the time of construction to induce steadier air within the tube, thereby reducing the effects of internal "seeing"); it pivots in declination within a wooden-braced latticework polar axis. The drive was originally a falling-weight one, long ago replaced by an electrical device that nevertheless retained a worm drive to the original, massive, six-foot-

Looking north through the latticework polar axle of the Great Northumberland Refractor to the observer's chair. Note the wooden square-sectioned tube with the finder mounted below.

diameter gear wheel that is fastened to the base of the polar axis.

In former days the dome was a wooden structure that rotated on twelve cannonballs! The observer was seated on a traveling couch that could move up and down, and rotate about a pivot at the center of the dome floor. From this vantage the astronomer could control the rotation of the dome, the slow motions of the telescope, and the tilt and height of the chair, all by means of an ingenious system of levers and ropes. These functions are still available, with the sole exception of the facility for dome rotation: alas, the flimsy old dome and its military ball bearings have long since been replaced by a sturdy and quite massive metal dome calculated to maintain firm muscle development in potential observers.

It was really at Cambridge that my own astronomical interests broadened. I was equally happy collecting faint planetary nebulae, star clusters, or galaxies with the fine 80x wide-field eyepiece, mapping the solar system at 220x, watching for the elusive minima of certain variable stars, or splitting close double stars at high magnification. I was in celestial clover. Early in the evening I would scurry back into town for dinner at the college, after having enjoyed a fascinating hour trying to grasp the elusive markings of Venus, or having timed some phenomena of the moons of Jupiter, or after having estimated from the globe of Saturn the position angles and separations of the satellites for subsequent identification. My appreciation of the planets really blossomed in this period. The Northumberland was a splendid telescope, despite the rather sad condition that one of the components of its objective was devitrifying (that is, the glass was reverting to a crystalline state), thereby producing much internal scattering of light and a softening of the image. It would reveal much subtle planetary coloration and the vaguest of features. Indeed the challenge was always to record on paper exactly what one's brain claimed to have seen. The first planetary drawing in my Cambridge observing log is of Saturn, in October, 1966. It looks like Jupiter, actually; just a flattened globe, crossed by shadowy bands. No rings. The earth intersected the plane of Saturn's rings three times in 1966, and my first views of the planet in a large telescope were more curious than revealing!

A quantum jump also occurred in my astronomy simply by virtue of being able to see fainter than, say, tenth magnitude, the limit imposed by the small telescopes I had hitherto used. I well remember silent March nights in the dark of the moon, when I would scour the galaxy fields of Coma, Virgo, and Leo in quest of faint NGC objects (the *"New" General Catalogue* of J. L. E. Dreyer, issued in 1888, that collected together all of the faint nebulae seen by the Herschels and their contemporaries into a single book). Since most of the objects for which I was searching lay within an area of only a few square degrees, there was little need to turn on the talkative drive. Therefore, it was dark and still — just the telescope and the observer. At times like those one could truly appreciate the romance and recapture the sense of discovery of the former great age of astronomy that was already advanced by the time Queen Victoria ascended to the throne of England. One felt close to the heavens, close to the thinking that motivated early pioneers. Indeed, many a quiet night as I sat high up on the observing chair, laboriously sketching faint fuzzy objects, I could almost sense that the ghost of a former avid user of the telescope was critically looking over my shoulder, hopefully in approval. Many are the things that go bump in the night around the Northumberland dome, perhaps some even attributable to the shade of Challis. Who was Challis, you ask? The answer to that question merits a short story.

James Challis was a Victorian astronomer who was responsible for a host of interesting astronomical inventions. However, in Cambridge he is best remembered for what he did not succeed in rather than for his many significant achievements. Challis was the astronomer that Neptune failed to find in 1846. Let me clarify my word play. When was Neptune "discovered?" In 1846. By whom? By Johann Galle and Heinrich d'Arrest of Berlin. How and why? Now the explanation becomes knotty and tangled with the personalities of the time. In the 1830's, only seven planets were known. The Astronomer Royal was Sir George Biddell Airy, apparently a somewhat testy and ruthless individual. Airy had been sent a paper by John Couch Adams, a young Cambridge mathematician, in which the latter predicted the existence of a planet beyond the orbit of Uranus, based upon purported perturbations of the motion of Uranus that were inexplicable solely in terms of the attractions of the then-known planets. Eventually Airy responded by making an irrelevant criticism of Adams' calculations. There the matter rested for a while. In France, the theoretician Urbain Le Verrier, by studying the motions of Saturn and Uranus, had also predicted the existence of a trans-Uranian body of large mass. Adams had requested that Airy relay the estimated approximate position of the new planet to Challis, who would be able to seek the body with the big Northumberland telescope. Airy was tardy in complying with Adams' request to the extent that the French, represented by Le Verrier, whose prediction was accepted and acknowledged by the Académie des Sciences, stole a march on the British.

Ultimately, Challis began to seek Neptune. However, to recognize such a planet he would have to find a body that was moving with respect to the background of fixed stars in the direction, and at the position, suggested by Adams. Nowadays we take for granted the existence of deep and detailed maps of the sky, but such charts only existed for a very limited portion of the sky in the 1840's. The charts were being constructed and published, section by section, by astronomers in Berlin from painstaking observations. The predicted position of Neptune lay on a chart not then published. Consequently, Challis was obliged to map tracts of the sky for himself rather than simply seek an interloper among a known pattern. On several occasions Challis had actually noted such an interloper and had sometimes even commented on his perception of a small disk. So, you see, Neptune failed to find Challis. The Berliners found the planet in 1846 close to the positions computed, independently, by Le Verrier and by Adams. There is a little-known ironic epilogue to this story. Actually, Neptune was located at its discovery on the overlap zone between a published sky chart available in the Cambridge University library, and the unpublished chart. Of course, Challis could not have known in advance the accurate location of the planet, hence he had no reason to hunt in the margin of the existing chart. A somewhat bittersweet story, I think you will concede.

One of my fondest amateur astronomical memories of Cambridge doesn't even involve a telescope — excluding clinging to the weathervane atop the old Greek-fronted building with one hand and grasping a huge pair of binoculars with the other in a fruitless attempt to catch Comet Daido-Fujikawa in 1970 as it grazed the sun. It was one of those occasions when a major meteor shower was due, the January Quadrantids, and several enthusiasts had arranged to gather at the Observatories for an all-night session. Perhaps a dozen of us arrived at about 9 p.m., replete with stopwatches, cameras, tape recorders, and deck chairs. It was, superfluous to note, cloudy. The throng lingered for about two hours and then dispersed. I had cycled most of the way home when I spotted an ever-increasing patch of cloudless sky. Wearily I retraced my tire tracks and found that two equally masochistic friends had done the same. The clearing was dramatic and in earnest. The three of us simply lay on our backs on the cold clammy ground and observed and recorded for ourselves. In four hours we saw 160 meteors altogether, including some quite spectacular ones. I shall leave the moral implicit!

A host of astronomical memories parades before me as I recall Cambridge: Saturn ringless; bright clouds on Mars; the central star of the Ring Nebula in Lyra (about magnitude 14.5 and glimpsable in the Northumberland); Pluto; predawn expeditions to catch shy Mercury before sunrise; the lacewings that lived within the Northumberland tube, sometimes emerging indignantly as one changed eyepieces; the hedgehogs, lurking outside the dome to trip up the unwary; Leonids, Perseids, Geminids recorded on thirty-five millimeter Tri-X emulsion; a supernova in an external galaxy.

The Observatories provided more than just the Northumberland. We had access to an equally fine and almost equally historic telescope: the Thorrowgood 8-inch refractor, equipped with a bifilar micrometer for double star work. I established an interesting intimacy with the Thorrowgood one summer when I took advantage of the relative absence of undergraduates to attempt to set up the micrometer and measure close double stars. The micrometer capably absorbed weekends, days, and nights, as I studied the mechanical construction of the device, calibrated the two screws, and developed techniques for measuring very close pairs. One of the most interesting problems was feeding a faint red glow into the field to illuminate the minute dark wires against the dark sky. I was able to measure separations between one and two seconds of arc fairly reproducibly, and to elongate pairs and estimate separations down to 0.4 second with difficulty. For such close work, involving at least one brightish star, it was often as accurate to estimate separations in terms of the spacing of stellar diffraction rings as to measure the pair directly with the wires. My "favorite" type of double star was undoubtedly the moderately close pair (less than ten seconds of arc) involving several magnitudes difference in brightness and strong color contrasts.

My most ambitious use of the micrometer was an attempt to perform absolute astrometry on Barnard's Star, the tenth magnitude red dwarf that is so close to us (six light years) that each year it moves some ten seconds of arc against the background stars. The idea was to use the micrometer and its rotating (position angle) stage to determine the directions and separations of a number of neighboring stars with respect to Barnard's Star. I made measurements each

The Thorrowgood refractor with its conventional equatorial head. Shafts with wooden handles are for clutches and slow-motion controls. At the base of the pier is the old falling-weight drive, now replaced by a tiny electric motor.

summer for three consecutive years. I can definitely assert that Barnard's Star moves perceptibly from year to year.

I had often wondered at the size of German battleships. I wondered no more after I encountered the huge, but excellent, ex-German-battleship binoculars provided for the amateurs at Cambridge. These are 10 x 80's; well, I did say they were huge. They had built-in filters and a supermassive tripod. They were perfect for brightish variable stars (like Nova Delphini 1967), asteroid hunting (I personally ran down Ceres, Juno, Pallas, and Vesta with them), and bright comets (oh, Bennett; you were gorgeous). The society would host open evenings for all interested members of the university, three evenings a week. The binoculars really pulled their weight — which was considerable — during those sessions. We could set up a couple of planets, or the moon and a planet, in the two big refractors and also provide rapid, dramatic stargazing through the binoculars. The evenings were extremely popular, even if I subsequently lost my voice after quelling the concomitant barrage of questions that accompanied a good night's viewing.

There were two other domes, basking in the sodium glow of nearby street lights and nibbled around by a local herd of sheep. One housed a 17-inch (corrector plate)/24-inch (primary mirror) Schmidt telescope, the other, a 36-inch reflector. The Schmidt did sterling work of an astrometric nature and was responsible for upgrading optical coordinates of quasars, for comparison with the highly accurate radio interferometric positions. It was also the observatory's standard instrument when worthy comets crossed the sky. I recall my one and only comet plate with the Schmidt, of Tago-Sato-Kosaka (1969g), taken in the winter of 1970 and showing a couple of widely separated tail streamers.

I never myself worked on the 36-inch, the principal professional workhorse of the Observatories. An effective technique had been evolved through the 1960's for extracting valuable spectroscopic information from stars bathed in the well-illuminated night of Cambridge. The procedure was always to compare one spectral line of interest with another piece of spectrum, close by in wavelength and observed simultaneously. The ratio of these two spectral chunks was effectively freed of terrestrial, in particular man-made, contamination. The telescope also wore a variety of other fascinating instruments, two of which I should like to mention. Think of observing a single star, let's say of sixth or seventh magnitude, in a fine refractor. Try to draw from your memory and experience the exact appearance of the Airy disk (the central circular core of the image) and the outer diffraction rings. Now let us pose some simple questions. What does this total diffraction pattern look like? Are the rings equally bright around their circumference and in comparison with their inner and outer neighbors? How does the seeing affect this pattern?

This problem was exceedingly fascinating to Dr. J. J. Linfoot, the resident optical and mathematical wizard of the Observatories; a short, taciturn, white-whiskered gentleman who always recalled a Herschel to my mind. Linfoot worried at a mathematical representation of the atmosphere that could model the seeing, and he made quite specific predictions as to what an extremely rapid snapshot, truly an instantaneous picture, of a star image would reveal. For example, the bubbling of the atmosphere would remove energy from the Airy disk and brighten the outer diffraction rings, but these would not appear as smooth, continuous, uniformly bright rings. Rather they would consist of a sequence of regions of varying brightness, one or two of which could momentarily rival the intensity of the Airy disk, hence their name — "rivals."

How could the theory be tested? During my sojourn as a graduate student in Cambridge, a new rapid image-tube system was nearing the day when it could be applied, in anger, to a telescope. This enabled extremely rapid exposures of bright stars to be obtained, which were recorded on a linear photographic emulsion (where the density of star images is proportional to the incident light intensity, as opposed to the usual photographic logarithmic characteristic curve) that could subsequently be traced very precisely with a microphotometer. A series of fast photographs revealed all the features predicted by Linfoot; a delightful accord and collaboration of theory and observation.

I vividly recall watching another intriguing device churning away (literally, as you will see) on the 36-inch. This was a very high speed photometer, designed to look for quite small-amplitude flickering in a host of interesting stars (for example, old novae like DQ Herculis with its seventy-six second periodicity) on a rapid time scale. One crisp night in the early winter of 1969, I packed up the Northumberland after several frosty hours of planetary sketching and drifted toward the dome of the 36-inch, whose slit was open and inviting. A continuous chattering was audible in the night calm, even from a considerable distance. Intrigued, I entered the dome and began to ascend the stair to the observing floor. Then it saw me: it was incredibly long, extremely swift, and pink. The snake lunged at me, at my face, level with the observing floor. I leapt back instinctively. Hearing my footsteps, an anguished voice called out, "Hey! could you catch that paper tape, please?" Tens of feet of pink snake, patterned with a sinister mottling, slid out of the dome toward my feet. In relief, I scooped up yards of the adventurous perforated tape and restored it to its rightful home, a vast tomb of a packing crate on the dome floor. The light intensity of a star was read by the rapid photometer and punched out at great velocity on paper tape (of all possible media!) for subsequent Fourier analysis on a large computer. I hate to think how many feet of pink tape were generated by twenty minutes of solid observation. The source that evening whose data went hurtling past me? The Crab pulsar.

3 THE ROYAL GREENWICH OBSERVATORY

It was my second year as a mathematician at Cambridge University. I had applied to the Royal Greenwich Observatory to spend a part of my summer that year working as a student with the resident astronomers on the south coast of England. Some dozen students were on the vacation course with me and, in retrospect, one can now see that about half of us continued with careers in astronomy. This Royal Observatory is beautifully situated, scenically that is, with offices in Herstmonceux Castle (reconstructed in the 19th century). The grounds encompass many acres of woodland, lake, and garden, and on a slight rise overlooking the castle live the domes.

The summer involved several formal lecture courses, an apprenticeship with one of the astronomers, and exposure to professional astronomy on several of the available instruments. The principal conglomeration of domes, the so-called Equatorial Group, includes two large refractors with 26- and 28-inch objectives, a 13-inch refractor, and two reflectors of aperture 30 and 36 inches. Separated from these domes by perhaps one hundred yards towers a large aluminum-sided cylinder accommodating Britain's largest optical telescope, the 98-inch Isaac Newton reflector. The height of this cylinder was necessitated partly by the not unusual desire to avoid disturbed air near the ground (like the Kitt Peak 158-inch telescope dome, Chapter 6), and partly by the thick white fogs that would often form yards high and deep over the marshy ground that lies between the castle and the domes. Supposedly each student was to spend at least one clear night on each telescope. I was responsible for drawing up a schedule for the students.

Herstmonceux Castle at dusk.

The Equatorial Group of domes at the Royal Greenwich Observatory. From left to right, these house the 26-inch refractor, 30-inch reflector, 13-inch refractor, 36-inch reflector; and 28-inch refractor.

The weather was sufficiently poor and erratic that even my own night on the Isaac Newton was completely cloudy! It was that sort of a summer, observationally.

As I look back on the period, there are really only three telescopes that either made a significant impression on me or with which I established any kind of rapport. They will not be predictable by aperture; indeed, two of them I have not even mentioned as yet. Inside the lovely wood-paneled dome of the 26-inch refractor sat a tall wooden tripod bearing a 5-inch refractor. Instantly it evoked memories of my beloved high school 3½-inch, and I obtained permission to heave, drag, or otherwise coax this instrument out of the dome to use it. It weighed as much as the massive brass-fitted tripod of high school days and was comparably long and unwieldy. But I enjoyed myself immensely with it. Since I had spent most of my life in the north of England and was rarely present in Cambridge (in the south) during the summer months, my knowledge of the southerly constellations was ragged and appalling. I began to work on this lack with the 5-inch and collected numbers of pretty globular clusters from the Messier catalogue.

Glancing through my copious observing notes for that summer, I am reminded of an anecdote that is appropriate to pass on. I was seeking Comet Honda (1968c) in a position described by a comet circular as "near to Capella." I dutifully pried the 5-inch out of the huge dome one evening and pointed, rather pessimistically, at Capella. There it was! Comet Honda — conspicuous, sharp, like a fan with curved edges and one side much brighter than the other. It was the easiest comet for which I have ever swept. The next night, too, I located it readily again close to bright Capella.

I began to try to calculate the time of perihelion passage. On the first night it had been north of Capella; on the second, south. My computations were totally out of step with the official estimates. I went back to the telescope a third night and located the object. I was about to sketch it again when some quirk made me sneeze. The telescope quivered. So did Capella, not unnaturally. However, "Comet Honda" quivered in such a manner that it approached and receded from Capella every couple of seconds. I had sketched and followed a ghost, an internal optical reflection in the telescope. The moral is that if this happens to you, or if you are observing, for example, Saturn and you can identify all but one satellite, tap the eyepiece and watch very carefully this unknown object relative to the bright planet (or bright star). I eventually found the real comet a week later, much fainter than my ghost had been, and was able to follow it over the next three months (with the aid of the ever-faithful Thorrowgood).

The 26-inch was an impressive gem of a telescope. None of us was permitted to use it, however, and we spent only one evening, as a group, peeping through it. I bear with me a vivid impression of the dense central region of the globular cluster M15 in Pegasus as seen through this giant lens. I was hoping for a glimpse of that rarity among rarities, a planetary nebula in a globular cluster. (No, you can't count NGC 2438 because M46, in which this lies, is an open cluster.) There is a planetary in M15, but it is of stellar appearance and quite faint.

Structurally the 26-inch was nostalgically reminiscent of the Northumberland. It too boasted a hollow polar axle constructed of beams in which swung the refractor tube on a stubby declination axis. The 26-inch, however, had one built-in device that was unique (actually, there may be one other) and which greatly appealed to me. The "comparison image micrometer" was an ingeniously fabricated mechanism that enabled a pair of artificial stars to be inserted into the field of view. One could vary the absolute brightness of the pair, the relative brightness of primary and companion, the position angle of the double, and even the separation. In order to discover the parameters of any celestial pair you needed only to match the characteristics of real and artificial pairs and read off, on the scales provided, position angle, magnitude difference, and separation. It was well-suited to work on difficult close doubles, where the null test involved in matching the two sets of images would lead to far more reliable results than would any attempt at direct measurement with a filar micrometer. But I could not get an opportunity to use the device, to my bitter disappointment.

The third telescope was one I used far more than any other that summer. It was a 6-inch refractor that did sterling service as a solar monitoring telescope and yielded uncounted numbers of photographs. The Greenwich sunspot index was defined on the basis of studies with this instrument at that time. I was granted permission to use it during the nights and found it to be a fine telescope for my eclectic interests in anything celestial, from planets to planetaries. It possessed a grand eyepiece for wide-field sweeping, and my fondest memories of the 6-inch are of the open clusters in Cassiopeia (picked off the *Atlas of the Heavens*).

There is a fine piece of old well-kept turf that surrounds the 6-inch dome. One afternoon I was talking with the astronomer responsible for maintaining the solar program. We were standing on the lawn by the little dome. I forget the original topic of conversation, but we ended up discussing rabbits and the threat they pose to a good piece of turf, constantly scampering about with their abrasive feet and emulating the subterranean tendencies of miners. Apparently this had also worried an earlier generation of astronomers. One man, now a famous theoretical astronomer, decided that it was possible to nullify this disastrous threat by suitable ingenuity. He carefully observed the routes taken by the leporine population and noted a small number of densely used paths that intersected here and there. He set up shotguns across the lawn, triggered by wires planted at these points of intersection. Several weeks passed; weeks punctuated by frequent bangs and by the creation of earthy craters in the once verdant lawn. Either the rabbits were totally annihilated by this artillery or their peers assured them of prompt burial; for no corpses were ever found. (Well, I did say the man was a theoretician, didn't I?)

It was also a summer for learning the philosophy of professional observers, a philosophy that was at times quite indistinguishable from cynicism. Sometimes we students would arrive early in the evening at the dome to which we had been allocated. The sky would be clear but there was no astronomer. After half an hour of exasperation we would telephone the relevant individual and protest the clarity of the Herstmonceux sky. Our words were often greeted by a comment to the effect that if we waited another half hour it would be cloudy. Such pessimism was not always appropriate. While we were awaiting the predicted clouds we would attempt to read by the light of glowworms. An even dozen sufficed to produce a fairly steady yellow-green lucency. Another pastime in which we indulged with enthusiasm was counting the death-traps scattered between the domes of the Equatorial Group. Lest you envision minefields, let me clarify. The domes are fairly evenly spaced, basically in two rows. There is a pleasant pond in the middle, and a series of paved areas, divided by low brick walls and linked by small staircases of several stone steps. In short, on a moonless night it would appear to be a relatively easy matter to find oneself in the pond, or lying on one's face at the foot of a stairway.

On a cloudy, partially moonlit night, the castle and its moat projected a somewhat eerie ambience. At such times it was easy to

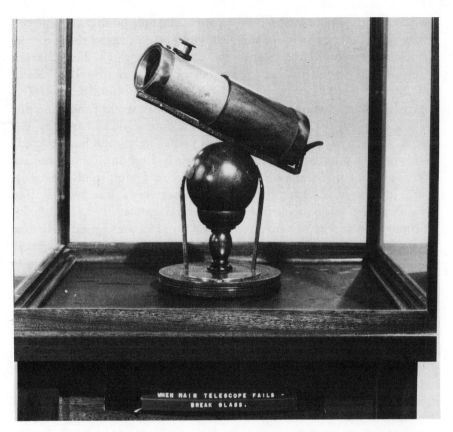

This model of Isaac Newton's famous reflector was encased inside the dome of the 98-inch. Note instructions! Photograph from the Royal Greenwich Observatory.

believe in the tales of miscellaneous specters that continued to haunt the castle and its grounds. There were Ladies in Grey, Girls in White, a Headless Knight on a bridge across the moat. I never saw any Unidentified Flitting Objects, however.

I have said almost nothing about the 98-inch, I realize, but I simply do not know the instrument. It was certainly a huge telescope and the largest I had ever seen when I first encountered it. The sheer magnitude of the volume enclosed by the dome was awesome. To contemplate the lowering of a 98-inch mirror through the giant doors in the observing floor, down into the aluminizing facility far below, was equally difficult. It had not been in operation long by that summer and it was, of necessity, still being debugged. Consequently, there were abundant tales of every mischievous trick perpetrated by the telescope. For example, there was allegedly a night when the prime focus cage was set up, the observer was ensconced therein, and the telescope refused to lower to the position in which this observer could quit the cage. Instead, it stayed up in the air, rotating slowly about its axis — definitely redolent of a medieval torture device. The internal dome structure was clearly incomplete at the time and remained so when the instrument was to be formally dedicated by the Queen. Security regulations demanded that a fire escape be provided. In response to this requirement, an extremely long and flimsy-looking ladder was set up spanning the gap between ground and observing floor. In case of an inferno, the Astronomer Royal was to lead Her Majesty to the safety of terra firma along this route.

Toward the end of my summer at RGO a plan was formulated that would definitely place astronomy once more among the pursuits of gentlemen. The Astronomer Royal lived in a wing of the castle and it was his intention not to have to leave the comfort of his home in order to observe. The 98-inch was to be equipped with a type of television sensor, coupled to a 10-inch finder, that would enable remote guiding. I do not know whether that degree of remoteness was ever attained, although conceptually it could have been. Remote operation is desirable for the largest telescopes, where access to an elevated Cassegrain cage, for example, is not always convenient during a night. Indeed, in 1979, California astronomers are thinking seriously of a 300-mile-long remote link to a future telescope in a potential very high and inhospitable mountain site.

The summer of 1968 was a vital one for determining my attitude to astronomy as a profession. It introduced me to large instruments, to formal astronomical education, to professional astronomers. It gave me clues as to the motivations behind the research concept and as to the design of experiments. It showed me that, if one were to be based in England, then one would have to rely on tapping sources of good weather in other countries. For the first time, the magical names of Palomar, Kitt Peak, Pretoria, Pic du Midi did not sound so impossibly remote as they once seemed. Ideas were beginning to crystallize in my mind. Now it was just a matter of carefully selecting the type of astronomy with which I would like to work.

MINNESOTA 4

In summer of 1969 I was graduated from Cambridge as a mathematician. I had already made my next decision — namely to pursue my astronomical interest. The fundamental issue was whether to become a theoretician, as befitted a mathematician, or an observer to appease my "amateur" heritage. To go into theory at Cambridge required another year of courses, a notoriously tough year even for good mathematicians, at the end of which there could be no guarantee of a spot in Fred Hoyle's then recently inaugurated Institute of Theoretical Astronomy. I pondered very carefully a statement made to me by the Astronomer Royal, with whom I had worked at Herstmonceux. He had advised me never to select an observational career just because a theoretical one seemed difficult. I had just completed Finals. It would be pleasant to dispense with formal courses and to enter the ranks of the observers. If I were to do that, which of the usual fields would I choose — optical or radio? In practice, a very good friend had just opted for the observational life and was about to be shipped off to America to learn the practical side of infrared astronomy. It intrigued me. Here was an observational field, recently established, needing new blood and enthusiasm. It would be like starting astronomy again, with fresh eyes on the universe. It also offered the merit of travel, to the United States. The immediacy lured me on. Even a lowly graduate student might have an impact on science through a new field.

I had made my decision, one frosty January morning as I was walking through the great courtyard of Trinity College, assailed by the

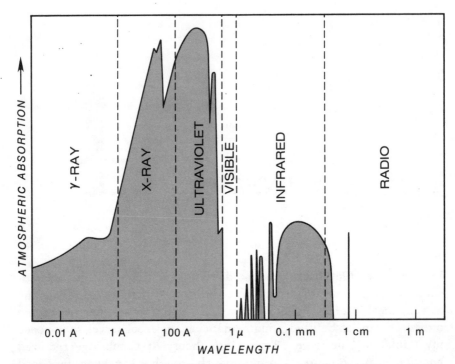

Very low dips in the shaded area indicate the wavelengths of "windows" in our atmosphere through which we may observe the universe. Along the bottom scale, from left to right, the abbreviations refer to angstroms, microns, millimeters, centimeters, and meters, respectively.

tintinnabulation of the clock tower. It had been a decision taken at the emotional level, requiring only time to become rationally firm. I did not relish the idea of spending perhaps a year or two finding an abstruse theoretical problem that no one had yet tackled. It might take a long time to find what had not already been done by the great father figures, like S. Chandrasekhar, or by the bustling modern school of quasar pundits. I would try for infrared!

I began work at the Cambridge Observatories, reading physics books and astronomical journals, essentially in isolation. When my friend returned from his second trip to the United States he gave me the benefit of his perspective on infrared. Who was studying what problems, and were the existing measurements reliable? I had, for various reasons, become interested in star formation and in the study of young objects. After all, stars are born in dark clouds where dust abounds, and the penetrating power of infrared is useful in these obscured environments. Moreover, the first bright objects in the infrared were being found, and R Monocerotis the bright pseudo-stellar tip of Hubble's Variable Nebula (NGC 2261), possessed a good deal of unexpected (at the time) long-wavelength radiation. The nebula looked like a young system and R Mon was somehow allied to the group of allegedly unevolved (pre-main-sequence) T Tauri variable stars. I decided to attempt a broadly based long-wavelength survey of optically visible young stars, once it came my turn to head for points West in quest of infrared telescopes.

It was the fall of 1970 when I met the United States for the first time. Culturally, the transition was not as abrupt as it might have been, for I moved to Minneapolis rather than to California. Nevertheless, there was much to absorb. More to the point, I met my first infrared equipment.

The optical window to which our eyes respond is an extremely thin slice of the total electromagnetic spectrum. As we move to longer wavelengths, not only do our eyes cease to be sensitive but photographic emulsions and image-tube phosphors also become useless. The infrared is a vast domain, spreading roughly from one micron (10,000 angstroms), the upper limit of photography, to about one millimeter, where attitudes and techniques more closely resemble those of radio astronomy than of infrared. This huge spectral expanse is severely cut up by numerous absorption bands due to gases in the terrestrial atmosphere, yielding opaque regions and more or less transparent "windows." Infrared astronomy is conducted in the windows between these opaque zones. Conservation of energy wreaks one more trick upon us. If there are spectral regions in which the atmosphere is absorbent, then the very gases themselves heat up by virtue of this absorption. In fact, the equilibrium temperature of the atmosphere is close to 300 degrees absolute (around room temperature) and that temperature assures peak emission in a prime infrared window, around ten microns in wavelength. Through this hot haystack the astronomer must look for his warm needle. The sky is never dark in the infrared; there are no "night" and "day" distinctions.

Infrared astronomy has evolved special techniques to deal with this bright background of radiation. Imagine that instead of squinting through a telescope with one eye at a star in a bright sky you used a pair of binoculars and kept both eyes open. This is not an ordinary pair of binoculars, however; the two glasses point at very slightly separated patches of sky. One eye perceives star and bright sky, the other sees just empty bright sky close to the star. Now, blink rapidly from one eye to the other and ask your brain to subtract what the left eye sees from what the right eye sees. In this manner the bright sky background could be canceled and the difference between the two eyes would be the star and, ideally, only the star. Essentially this is the sky-chopping technique so vital to infrared astronomy. In practice, one detector is used (rather than two, possibly unmatched, eyes) and this is "chopped" rapidly (ten to thirty times a second, typically) between two areas of sky.

The essence of an infrared system is a detector that responds to incoming radiation and generates electrical signals. These detectors are tiny in physical dimension, often less than half a millimeter square, and are always operated at extremely low temperatures, close to absolute zero. The justification for this cryogenic (low temperature) approach is that the celestial signals are so weak that all sources of noise internal to the detector and its associated electronics must be quieted. At the microscopic level, the material from which the sensors are constructed is a lattice of atoms, mostly of germanium or some other semiconductor, bathed in a fog of electrons. This lattice and its electrons are constantly quivering very slightly, and these vibrations are in response to the temperature of the material. The colder the detector, the less the quivering and the less the internal noise due to jostling of electrons and atoms.

I also met liquid helium for the first time. Infrared detectors are housed inside vacuum-tight thermos containers called dewars. In order to cool these vessels to extreme temperatures, they are constructed essentially of two or three nested containers that hold one or two different cryogens, or cold liquids. Typically, liquid nitrogen is fed into the dewar and this drops the temperature to seventy-seven degrees above absolute zero. Then the innermost copper flask is fed liquid helium and alchemy is brought into play. Liquid nitrogen is a well-behaved liquid, rather like water or root beer. You can pour it out of a flask and it will splash (and, admittedly, sizzle) around the floor. It will prick your hands with cold if it splashes onto you. By comparison, liquid helium is mysterious. It never makes it out of the neck of the flask. Invert the container and all that emerges is a cool, thick, white plume as the fragile liquid encounters the warm air in its environment. Transferring liquid helium from a large storage vessel into a small dewar that will sit on the back of a telescope is definitely a

Liquid helium is transferred through the thin, vacuum-walled, inverted U-tube from a container in the box at left to the infrared cryostat. At the bottom of the latter, the guiding eyepiece can be seen.

black art, although one that can be learned with experience. There is one nasty hazard of which one should be aware. It is possible, if one is careless, to suck warm air into a dewar containing liquid helium. This freezes out a solid plug consisting of ice and frozen air in the slender neck of the dewar. Helium that is undergoing the transition from liquid to gaseous phase expands, volume for volume by 10,000-fold at sea level. A plugged neck over liquid helium definitely does not make for soothing contemplation. If the pressure of gas behind the obstruction is not relieved by the violent expedient of ramming a warm metal rod down the neck, disastrous consequences can result. For example, the carefully crafted infrared system can be liberally distributed over the ceiling and walls of its immediate environment.

Enough of the perils and horror stories of infrared technology. The University of Minnesota at Minneapolis was home to one of the very few groups undertaking the development of infrared equipment for application to astronomy. The university owns a beautiful 30-inch telescope of optical quality, set along the leafy banks of the St. Croix river. Driving back from the observatory on early fall mornings is an unforgettable experience. Tree after tree is coated with red, gold, and orange, contrasting sharply with the deep blue skies.

It was a drive of about thirty miles from campus to the O'Brien Observatory, close enough not to represent an expedition, yet far enough to necessitate some contemplation of the weather prospects. In the office occupied by the astronomy graduate students was an old radio that could be tuned to receive Federal Aviation Administration weather reports. "Taking the weather" became a ritual carried out at least thrice daily. It involved noting the cloud conditions (at three different altitudes), temperature, dew point, and wind velocity on a chart showing the locations of major cities in Minnesota and the Dakotas. It taught me how precise the art of weather forecasting could be in an inland environment from which the oncoming weather could be sighted when as far off as a thousand miles, in contrast to the British Isles, where microclimates are involved, so small is the scale of the country by comparison with the scale of weather systems.

On the basis of the weather, we would decide by late afternoon if it was worthwhile to trek out to the 30-inch. If so, we would descend to the basement of the physics department to collect an infrared detector — to me a wonderland of cast-off pieces of televisions, rockets, balloons, previous generations of infrared equipment — and piles of miscellaneous unclassifiable instruments. In this room and the nearby machine shop were conceived and built the beautiful infrared devices that we took to various telescopes, in Minnesota and Arizona. Like bony skeletons the older cryostats hung neglected on a wall, ready to be resurrected or cannibalized as the need arose. These dewars were cylinders about a foot tall and six inches in diameter. They consisted of an inner copper flask, for liquid helium, and an outer vacuum jacket, to keep at bay the hot outside world. Inside there were mounting plates on the cold copper surface for the tiny germanium bolometers, less than a millimeter square. It required a high-power binocular microscope to perform the operation of connecting tiny wires to opposite edges of the bolometers — wires that had to be cleverly designed to provide mechanical support for the detectors, their thermal connection to the cold bottom surface of the helium vessel, and the electrical route for signals generated by the bolometer when exposed to infrared radiation.

Miracles of micro-engineering regularly occurred in this basement. There were usually only one or two dewars that were telescope-ready, distinguished by the appellatives "three-banger" or later, "ten-banger," which numbers denoted the selection of broadband filters available inside the cryostats. (Many are the airline passengers who have sat next to me on flights to Arizona and who have suspiciously eyed the wooden box stowed neatly under the seat before me. On the side of the box would appear, in large untidy letters, "12-banger No. 1." I could understand their consternation . . . .) From the outside world, even to maneuver a change of filter deep within the cold gut of a dewar required the utmost design ingenuity. Only briefly could such a threatening thermal pathway be established, and then it had to be severed after the desired filter had been placed over the detector. There were screws and worms and levers and spur gears . . . I could digress at length on the vagaries of establishing a solid high vacuum

Inside a four-banger. The spun copper sphere of this early infrared cryostat is a radiation shield inside of which is a smaller sphere to contain liquid helium. The wire on the left is a filter changer. This cryostat is upside down, and the tweezers are probing near the detector. The beam enters through the bottom of this dewar (at the top of this picture).

inside a dewar; on the dewars that sometimes just didn't hold a vacuum; on the rubber O-ring seals that failed because of tiny particles of grit pinned between them and the adjacent metal surfaces; on the horrors of arriving at the telescope only to find that either an electrical connection to the bolometer had broken, or one of the nylon rigidizing struts that delicately maintained the internal optical alignment of the helium vessel had snapped. There were times of high frustration, but they were also times of learning — appreciation of design and of physical principles, hitherto exemplified only in books in my previous career as a mathematician. But we were on our way to the telescope.

We could drive off after effecting the transfer of liquid helium into the cryostat and after boxing the dewar in its individual carrying case. Minnesota was an ideal infrared site in the winter; it was extremely dry, especially when a cold snap arrived, bringing daytime highs of ten and twenty degrees below zero. But somehow we would arrive at the observatory, motor up the snowy driveway, and enter the dome. Inside we were self-sufficient: there was a kitchen area, a room with bunks, radio, television, refrigerator, and a small amount of backup electronics. The dome was thermally isolated from the living accommodations by an unheated passageway along which there was a temperature gradient of some hundred degrees in about thirty feet. The telescope had "real" optics — a high quality primary mirror — and was solidly built. Its drives and positional encoders made it a high-speed slewing instrument that could also point accurately when operated from the warm control room. Of course, somebody always ended up with the somewhat undesirable task of going into the dome to change filters between observations and to check that the edge of the dome's slit was not intruding into the infrared beam. (Domes are warm, even at twenty below zero, and produce enough radiation to swamp all but the strongest celestial infrared sources.)

In the early days of infrared, the ability to set the telescope accurately and blindly was crucial. The brightest calibration sources were the planets, especially Mercury, whose proximity to the sun made it always a challenge to the eye with a finder telescope. The obvious sources to investigate at longer wavelengths were known from an early near-infrared (two-micron) sky survey carried out at Caltech. One knew the locations of these sources with some precision. Often there were no optical counterparts, and finding charts were useless. Deadsetting and scanning the telescope were the modes of searching.

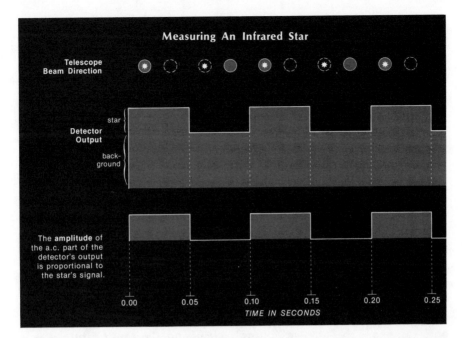

To distinguish between source and background radiation, infrared telescopes have wobbling secondary mirrors that cast a photometer's field of view ("beam") back and forth between the star being measured and clear sky. In the schematic representation above, the telescope's beam is placed first on the star and then away from it 20 times a second, resulting in the square-wave detector output shown. The a.c. part of the signal, which is proportional to the star's radiation, is integrated for as long as necessary to distinguish it from random noise; then the telescope is moved so that the photometer samples the sky on the opposite side of the source to eliminate background gradients.

There was little ambiguity in such a method, for the signals were giant, and one could almost feel for the bolometer, cringing in protest at the blast of photons! An auxiliary tool that we at Minnesota always used was a chart recorder. This would provide a continuous paper record of the signals from the detector, at all wavelengths. You could learn by careful appraisal of the chart if a source was well-centered in the infrared beam or if there were clouds about, recognizable by the hideously elevated level of background noise at the longest wavelengths. Also, it would tell if one had found a weak source, by the characteristic pattern of first right and then left deflections as the signal was switched from one beam to the next, a sky-chopping technique designed to remove not only the giant infrared-bright sky background but also gradients in this background. The ability to dead-set the telescope also made possible the twenty-four hour observing schedule, and sometimes we would have both day and nighttime observers.

My initial exposure to the cold of a Minnesotan winter was illuminating. I recall flesh freezing to the cold metal parts of the telescope; my moustache causing my face to be ice-bound to the metal stage that housed the guiding eyepiece (when it was necessary to use this); motors struggling to rotate the dome, even with graphite lubricant; small bulbs that died in the low temperatures; the first moment I started to believe the story about carrying words indoors to thaw them out. I also recall gorgeous auroras; sunrises so sparkling clean and daubed with pure color that you could cry just to experience them; flocks of geese flying silently over the observatory and creating havoc with their infrared radiation as they clipped the beams. I used to wonder how I would fare if I were trapped by a snowstorm at the observatory. In reality this only befell me when observing in Arizona (Chapter 5) and in Minneapolis itself.

One morning, as November was coming of age, I walked as usual onto campus. The sky was bitingly clear and the temperature was a windless five above. I decided that on the morrow I might need to don my giant down parka, until then reserved for observing sessions. It began to snow late that morning. It never stopped until five o'clock in the afternoon. I left campus at six. There were no sidewalks. These had been replaced by small mountain ranges of snow, plowed back from the roadways. The wind was gusting to forty miles per hour, and it began to snow again as I set off across a small snowy field about a quarter mile from the house I was then living in, just off campus. Halfway across the field, as I floundered shin-deep through the snow, the wind became turbulent and brisker than ever. I experienced my first whiteout. It was incredible. I could have died of exposure only a hundred yards from houses. There was no horizon, no direction; the wind spun every which way, and I was lost. Even my footsteps had been eradicated in my wake. But there were young stars to be measured in the infrared; the storm abated, and I moved onward in approximately the right direction.

I had decided to study star formation by observing T Tauri stars. These irregular variables form a class of several hundred known objects, the brightest dozen or so being visually quite bright (magnitudes from nine to twelve). They were just detectable at the longer infrared wavelengths, using a then state-of-the-art system on the 30-inch telescope. I would need something with a larger aperture to make a sufficient number of observations of different stars to seek patterns of behavior. There was the option to apply for time on the 50-inch telescope at Kitt Peak National Observatory (see Chapter 6). However, my arrival in America in late 1970 was at a highly serendipitous time.

5 MOUNT LEMMON

The universities of Minnesota and of California at San Diego, in conjunction with funding provided by the Science Research Council of Britain, were in 1970 ready to set up a 60-inch telescope, dedicated to infrared work, in a high mountain site in Arizona. Shortly after the first cold tingles of winter I left Minneapolis for Tucson, Arizona, driving with a colleague in a truck purchased especially for the new observatory (and in which I learned to drive on the "wrong" side of the road). Several of us arrived on Mt. Lemmon, the 9,200 foot summit of the Catalina Mountains just to the northeast of Tucson, on December 5, 1970. It was gray and overcast. We drove onto the then deserted U. S. Air Force base that until fall of 1970 had housed the 647th Radar Squadron. No dome greeted my eyes, just a two-foot-high circular concrete structure on top of which the walls of the dome were beginning to be raised. The air was cool and thin. Plans had been laid and correct phasing was important. The mirror had to be collected from its grinding and polishing site, taken to Kitt Peak where it could be aluminized in one of the large coating chambers, and returned to the Catalinas. During this period, the walls of the dome were to be completed, and pieces of the telescope were due to arrive. The telescope would be assembled in the finished building, the mirror would be lowered through the open roof, and then segments of the dome that had hopefully arrived would be erected on top.

I accompanied the observatory director, Nick Woolf from Minnesota, on the mirror expedition. We found this oddly sectioned hunk of

The Mt. Lemmon 60-inch metal mirror prior to aluminizing. The canogen coating and aluminum alloy are already optically quite reflective. Note the thin rim and thick central boss of the mirror.

aluminum hanging from a rusty central hook (it had a Cassegrain hole), in a garage of the optician. It was an all-metal, mostly aluminum alloy, mirror with a central boss about fourteen inches thick, tapering to a rim just over one and a half inches thick. It looked spectacularly shiny even then, consisting as it did of two coats of canogen (electroless nickel: the first coat did not prove entirely satisfactory) on top of the alloy. The infrared radiation would be reflected from these canogen layers, but an overcoat of aluminum would greatly improve the optical reflectivity. The mirror had crisscrossed America in a wooden crate on its travels to be stress-relieved, ground, and polished. It had previously traveled in style. After twenty minutes of demolition, we had removed enough of the wooden crate that had been its former home so that it was possible to squeeze the mirror into the bed of our pickup truck. We covered it with a sheet of plastic and began the nerve-wracking sixty mile trip up to Kitt Peak, as the $30,000 mirror jostled behind us.

I recall our arrival on the peak, which was so full of telescopes that I was quite bewildered. We drove our little truck with the 60-inch mirror to the entrance of the aluminizing building and found another truck already there. This was an elegant white-painted semi, belonging the the University of Arizona, from whose capacious interior I expected perhaps a 100-inch mirror to emerge. It did not. On the spacious bed of the vehicle lay a tiny crate containing a 40-inch pyrex mirror. This diminutive creature preceded ours into the chamber. I recall peering through the inspection window at the mirror before they flashed the aluminum coating at it, then trying to peer through the window after flashing, only to see my face staring back from the coated window. It was an unsuccessful attempt. A tiny droplet of water had run down the surface and streaked the aluminum. The man who had brought the mirror was moist-eyed as he drove away with his deformed charge.

Then it was our turn. With all the trepidation of expectant fathers, we watched the mirror mounted on what resembled an old bed frame, which disappeared into the chamber. For hours the vacuum pumps spat and gurgled, draining every unwanted molecule of air and water from the mirror and chamber. It was a beautiful coating, administered by professionals; in fact, they said it was the best they had achieved. We squeezed the mirror back into our truck, covered it with the wrinkled sheet of none-too-clean plastic, and drove off, leaving the coating men tearing their hair in anguish. The hardest part of the journey was the ascent of the Catalinas. We drove back from Kitt Peak on a Sunday and the mountains were awash with tourists, gazing at the valley, hiking by the road, and driving with chaotic abandon. It was stop and go and grimace traffic for forty miles, but we got the mirror up to the summit intact and untouched; if nothing else, the slow trip helped make it a safer one for the mirror.

Unfortunately, the various time schedules did not mesh as planned. The mirror waited. We all waited. The dome had arrived, but the telescope was delayed. We decided to go ahead. Erecting the dome was a fascinating procedure, as each segment of the final hemispherical structure was slid into its neighbor and pieces of bracing pipe were affixed for rigidity in incomplete circles around the fledgling dome. A small crane arrived to haul the three shutter sections up to their tracks. The telescope arrived — an incredible blue and white giant erector set of legs, cubes, and girders on the long bed of an articulated truck. Then a monster vehicle trudged painfully up the road from Tucson, bearing a crane with an eighty-foot jib: eight hours at five miles per hour equals forty miles. I pitied the tourist traffic that afternoon. This crane was used to unload the telescope components in what seemed to be a matter of minutes. The delicate part of the operation now was to maneuver this crane so as to lower the partially assembled pieces of telescope into the dome, through the

Into the aluminizing chamber in a vertical position.

The Ash Dome going up, leaf by leaf.

The arrival of the Astro Mechanics lightweight telescope.

seven-foot-wide slit. By that evening the telescope stood in the dome, naked and mirrorless. The worst was yet to come.

The crew building the base of the dome was far behind schedule. They had not yet constructed the floor of the dome; in fact, I had to build my own temporary floor that Christmas. Breathlessly, the mirror was lowered into the dome, past the telescope tube to the ground, three feet below the future floor level. Somebody rushed off to Tucson for another chain winch. It took eight of us over an hour with two hand winches, connected first in series and then in parallel, to raise the gleaming mirror some nine feet to the back of the telescope and to bolt it in place on its backplate. Twelve of us slept over at the observatory that night. I know, I cooked a dozen TV dinners. There were dome men, crane men, telescope men, plumbers, electricians, astronomers.

In retrospect, it seems amazing that on December 21, 1970, we obtained the first useful infrared data with the new telescope. We had only one operating detector system, an infrared spectrometer; the data were printed out on a roll of paper tape in a room furnished with a single wooden packing crate that served as table, chair, and cupboard. The telescope was clamped and released by bicycle chain affairs, and slewed by hand as fast as one could heave upon it; its pointing was with reference to setting circles. But the Mt. Lemmon infrared observatory was on the air and it became an impressive science machine and generator of PhD theses.

A contributory factor to its production of young astronomers was its comparatively poor images, limited by the quality of the metal primary mirror. I often feel that one of the cruelest things I had to say of the telescope was, when looking at Saturn, I stated that I had seen more planetary detail with my beloved 8-inch refractor at Cambridge. This was true, of course, but that had been the standard ethic about infrared telescopes. Since the wavelengths are so much longer than optical ones (by a factor of twenty, typically), one should be able to tolerate a much poorer image quality than an optical astronomer. To employ a cliché, it's not as simple as that.

Recall the fact that the infrared sky is always a "daytime" sky (Chapter 4) and it becomes clear that while chopping should, in principle, combat the background problem there may be slight gradients in the sky. Two adjacent areas on the sky correspond to two cones of reception "squirting out" from the detector through the atmosphere. Suppose the chunks of hot atmospheric gases (most particularly water vapor) along these two cones are not quite matched, then the subtraction inherent in chopping will not exactly cancel the background. Any imbalance between the two beams shows as a source of "noise" that tends to hide the signals from space. If the two beams are very close together on the sky, then we have matched the cones as well as we can. Nevertheless there can be spatial gradients in the atmospheric gases. It would also be desirable to narrow down our cone of reception so as to include the star and as little as possible of the surrounding sky. What happens if the seeing (the dancing and boiling motions of the star image) is very good but our mirror does not permit us to perceive images as small as the seeing disc, due to inherent optical blurring in the telescope optics? The answer is clear: we are forced to retain larger cones of reception than we would like. During the early years of Mt. Lemmon, the graduate students almost had the run of the observatory, for the interests of faculty members turned to programs requiring great sensitivity, hence very small beam sizes and the use of other telescopes. In 1975 a new mirror replaced the old metal one, an ingenious piece of construction that merits a digression.

Let us imagine ourselves grinding a metal mirror. Now, how does

Looking at the primary. Note the Cassegrain hole above center and also the extremely thin spider vanes that hold the secondary (reflection at bottom of picture). The spider is an infrared-optimized design.

metal differ from glass? It really ought not to differ much, provided that we have taken care to relieve internal stresses in the metal by thermally cycling the blank from cold to hot and back again. When we are grinding the mirror we attach it by several bolts to our machine, whereas a glass blank can lie on a suspension. When we mount it on the telescope, however, it is first attached by heavy bolts to a plate which is then connected to the telescope tube, unlike the complex floating suspension of large mirror cells. A different set of stresses exists on the telescope compared with the stresses during the grinding and polishing stages. The question uppermost in the minds of metal-mirror makers in the early 1970's was how to lose these stresses that, on the telescope, communicated themselves to the front reflecting surface of the mirror and caused unplanned distortions. It was an Air Force optical engineer who first solved the problem.

On the disused base of an old radome, another lightweight 60-inch telescope had been erected on Mt. Lemmon, to be used for lunar laser ranging. The first attempt at a metal mirror for this instrument yielded quite intolerable images. The second attempt embodied a new design. The mirror was constructed of two components: the optical surface was a thin layer of Cer-Vit, one of the low expansion ceramics that is replacing pyrex in modern optical design for reflectors. This layer, five inches thick at its center, sat on a Teflon pad and was literally epoxied onto a stainless steel mounting bush some eight inches deep. In section, the two-tone mirror was almost identical to the all-metal one; in weight it was similar, about 750 pounds, so the same lightweight telescope design was adequate. But now, all the stresses induced when the mounting bolts were tightened into the steel boss were lost at the soft interface between ceramic and steel.

You may ask, did it work? It was a clear night with excellent seeing, much better than one arc second. In great excitement an Air Force gentleman inquired whether I would like to give him my opinion about the image quality of his new mirror. Indeed I would, and we hiked up to the summit of Mt. Lemmon, into the dome stacked high with the huge laser system and wrapped in serpents of water-cooling for this beast. The moon was up, and I chose a series of tests that would put the mirror through its paces. The only one that I remember vividly now was a view of the Alpine Valley. I had never, and never have since, seen the slender crack running along its floor so clearly through a telescope; only half a mile across in sections, this indentation stood out clearly as a great wrinkle on the lunar surface. Yes, the two-component mirror works! By fall of 1975 the Minnesota telescope boasted a new Cer-Vit/steel primary and simultaneously ceased to be an instrument dedicated (or constrained) to infrared work. It has been used for optical photometry and spectroscopy, polarimetry, and submillimeter work, in addition to its traditional diet of infrared photons.

Let us plan an observing trip to Mt. Lemmon. First we must propose a program, send it to the director in Minnesota some three months before the desired date, and request time in the next quarter. If the program has merit, time will be granted. We fly into Tucson, meet the outgoing observers, and drive up the Catalinas in the late afternoon sunshine, collecting a week's worth of groceries en route. The air is thin. We carry our suitcases slowly into the spacious, comfortable, living quarters. One of us (there are usually only two observers at a time on the telescope) prepares a rapid simple meal, while the other goes into the dome adjacent to the dormitory to check out the telescope. Typically, two consecutive groups use very different instrumentation, and one has physically to mount the

The Cassegrain backplate of the 60-inch, with T-shaped photometer and guiding optics. Dewars mount onto the left arm of the T, parallel to the long dark pipe.

relevant piece at the focus. All its mechanical and electrical connections must be established. Any cryogenic preliminaries must be undertaken; it may require several hours of cooling to obtain the desired, extremely cold, and stable, operating point for some detectors. Once the instrument is ready, the computer can be awakened.

Virtually all pieces of equipment that are in frequent use at Mt. Lemmon are conceptually similar. The electrical signals from the detector are passed to a preamplifier, often inside the cold vessel to reduce thermal and electronic noise, or certainly physically very close to the detector, to minimize interference and loss of signal strength along extended cables. The signal is then fed to an amplifier which is capable of isolating from the spectrum of signals received from the preamplifier precisely the one frequency germane to the infrared signal, namely that at the chopping frequency (now usually defined by rocking the small secondary mirror from side to side.) These detected and amplified signals are fed to an electronic integration system, thence to a small computer which eats the individual signals and processes them statistically, updating and displaying the current result of combining all the previous signals. Data can be stored on magnetic tape, which is vital for spectrophotometry when hundreds of numbers are generated in a single night, or simply printed onto paper tape, the usual procedure for photometry when the data rate is much slower. We recheck the detector, open the slit to release warm turbulent air trapped inside the dome, and eat our steak (Arizona beef is excellent!).

As the sun sets, we suit up warmly and set about defining the beam of the system; in other words, we ensure that when we bring a star to a fiducial mark in the field of an optical guiding eyepiece its infrared signal falls simultaneously squarely on the detector. It will be necessary to move the secondary mirror considerably to establish the focus since we have put a new instrument on the telescope. In about an hour we are ready to observe, if all goes well. The program is already planned. We check the sky for clouds: remember, clouds both attenuate (absorb) and radiate (emit because they are warm) infrared; they are thus doubly a nuisance. The rest is a matter of evolving an efficient routine of setting up the stars (the telescope now has digital shaft encoders that read out right ascension and declination), locating and centering objects of interest, and accumulating enough data on each. Communication is by FM radio between the person in the

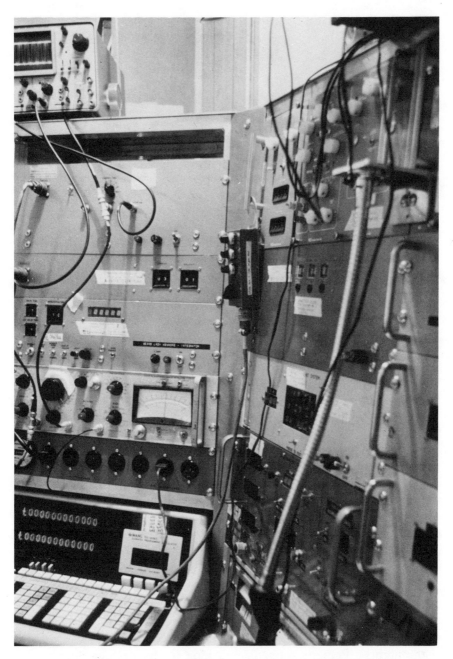

The heated control room on Mt. Lemmon. At bottom left is the Wang computer and at top left the oscilloscope that monitors the back-and-forth chopping of the secondary mirror.

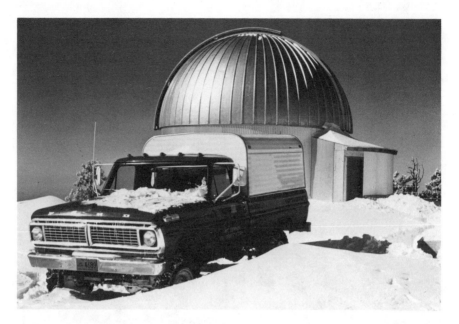

A typical Catalina Mountains blizzard; afterwards, the sky becomes very clear.

dome, who guides during the integration and sets up each object, and the person monitoring the data in the warm adjacent computer room. In case of peculiar data or deteriorating sky conditions, this dialogue is essential. It also serves to keep both people awake. By sunrise the telescope is put to bed, followed shortly by the observers.

By about 1 or 2 p.m. we are awake, examining the previous night's data, planning tonight's program, defrosting a chicken, playing badminton in the observatory's beautiful gymnasium, and establishing the cryogenic operating point for the detector again. One hopes for a week of clear nights, but rare is the trip on which this now occurs. In fact, looking back over the past eight years of statistics on telescope use reveals a disconcerting trend. Mt. Lemmon is normally open for astronomy three seasons of the year, being shut down for the summer monsoon in July, August, and September. In 1970-71 the nine months of observing provided data on sixty-five percent of the total number of nights. This percentage has declined year by year, and currently one expects to gather useful data on only about forty percent of the nights. Perhaps water is returning to the Arizona desert . . . . However, year by year the telescope and data room have been better instrumented, making for smoother, easier, and more efficient operation. There is even a sauna now. The attitude that underlay the establishment of the observatory was always that, at a remote site, the observers should be made as comfortable as possible in order to insure the best work. The laboratory and workshop are always well equipped. This is an ideal that I feel other institutions should emulate.

At 9,000 feet there can be a substantial amount of snowfall in the winter months. I recall being trapped on the mountain for four days one Christmas, after a blizzard deposited four feet of snow. If it is necessary to choose a high dry site for infrared work, why, you may wonder, was a relatively low peak chosen? There are higher and perhaps drier Arizonan mountains, but none is so accessible as Mt. Lemmon. It has a paved road, superbly engineered (it took over a decade of blasting to cut it) and quite well maintained by the state, rising to a ski lodge, where the prevailing attitude toward snowfall is generally at variance with that of astronomers. From the lodge, a three-mile ex-military road snakes slowly up the large, flattish summit area. I learned to drive in snow on this road, a sobering experience; I also learned to appreciate the powerful four-wheel-drive 396-cubic-inch Ford truck. The truck and an expensive microwave-link telephone are vital lifelines to the site, and they diminish in some measure its remoteness. In snow one must insure that the wind, that fickle prober of domes, does not force powder through the slit, and that any snow that does intrude does not fall upon vulnerable electronics nor upon exposed optics, such as inside the dewcap of the fine 6-inch refracting finder, a telescope that my "amateur ego" covets. In rain these same considerations apply, along with the necessity to climb among the rafters of the dormitory building with a bucket in each hand, to prevent rivulets from leaking through the ceilings of bedrooms, library, and living room. Mt. Lemmon has been a second home to me; since I first moved there in 1970, I have spent about an entire year of days and nights at the telescope and on site. It has been a delight and a privilege.

There are other nonastronomical benefits to working in the Catalinas. The drive from Tucson, at 2,700 feet above sea level, to the summit carries one through a host of different microclimes. The desert is beautiful, and the ascent presents an ever-changing balance of cactus species, until the Coronado Forest asserts itself and oak and juniper give way to pine. Life clusters at the foot of the range: snakes, scorpions, and tarantulas (occasionally glimpsed crossing the highway like dismembered black hands). On top there are hawks, finches, jays, raccoons, rabbits, deer, and giant moths. This leads to a quasi-astronomical digression.

When we first reached the mountain in 1970, the gymnasium was knee-deep in dead large moths. They have lingered ever since. To say that they have been obtrusive is to be guilty of gross understatement. I recall May, 1973, when even for Tucson there was an unseasonally warm spring of 105 degrees Fahrenheit in the valley, and the desert was awash with suffocating hot odors of wild herbs and flowers. On Mt. Lemmon, the moths were hatching with a vengeance. Their eggs were laid behind the secondary mirror, in electronics racks, throughout the dormitory, in the data room, below carpets — in every nook and cranny into which their athletic forebears could insinuate their brown papery bodies. We encountered a strange problem of "spikes" on the chart recorder that runs constantly in parallel with the computer and integrator. I sat in the dome one night with a powerful flashlight, while my assistant in the data room yelled "spike" every time the detector registered some enormous, rapid change of signal that wiped out that particular integration period. I depressed the switch on the flashlight. Moths would hover, snared in the beam between the secondary and primary mirrors, or peeping shyly down the tube, or flitting past the opened dome slit. We lost some thirty percent of our integration time at ten microns, a wavelength highly sensitive to stray radiation from ambient temperature surroundings. We did determine the temperature of moths with some precision.

The ultimate moth feat befell me in the middle of an April night as I was guiding on the dazzling glare of Vega. A strange snaky tendril was flapping about over the monster star image, and another joined it. I removed the guiding eyepiece and found a giant moth sleepily crawling inside the optics. It had flown into the Cassegrain hole of the primary, traveled down three feet of wide tubing, avoided certain death in a chopping mirror arrangement, bypassed a guiding beamsplitter that directed infrared to the detector and light to my eye, and come to rest peacefully an inch from my left eye on the big front component of the Erfle eyepiece. I know it must be a good story, for shortly afterward someone on Kitt Peak told me the tale with obvious relish, saying it had actually befallen some unnamed astronomer on Mt. Lemmon.

It would be impossible for me to summarize briefly the range of research that I have pursued at Mt. Lemmon. In 1972 and 1973 I shelved my preconceived observing programs and embarked, with a British colleague then visiting Berkeley, upon a series of studies that are only now nearing completion. Upon rereading our earlier papers I am aware of the way that a scientific paper is expected to proceed, in a rational chronological sequence that relates observations to thoughts. This is frequently a misrepresentation of the way that discoveries are made. They are not necessarily dreamed up in air-conditioned offices by pure logic, but can arise (if one is an observer) in the 2 a.m. cold of a winter's night, at the telescope, looking at some object for intuitive rather than rational motives. From such an intuitive discovery did my colleague and I work our way, ever more logically, through the infrared properties of a host of hot stars, as opposed to the cool stars ever popular in the infrared. In several places we found evidence for the formation of cool dust particles in the hot gaseous winds that emerge from these extremely hot stars, in spite of theoretical contentions that such environments were too hostile to the survival of dust.

I continued my studies of young stars, begun in my PhD days at Mt. Lemmon. Whenever a technological breakthrough enabled a new machine to be devised, I would eagerly await my chance to use such a machine at Mt. Lemmon on my favorite young objects. Thus it was that in 1974 I discovered remarkably pure water ice in the cloudy medium enshrouding one especially interesting young star. A friend at San Diego had built a new spectrometer that incorporated a generation of supersensitive infrared detectors, and I was privileged to use this device with him at Mt. Lemmon. I was seeking molecular fingerprints in the spectra of T Tauri variable stars in the two to four micron region. I examined a few potential candidates but saw nothing obvious. But it is necessary to process this type of raw data considerably to extract terrestrial absorption features; furthermore, ice occurs in a somewhat messy part of the atmospheric spectrum.

On my return to Berkeley I noted such a deep step cut out of the spectrum of one star that I could scarcely believe I had found ice. I sought some error in my work, unsuccessfully. Two months elapsed until I was rescheduled at Mt. Lemmon with the spectrometer. It was a snowy week. Five nights passed, incredibly snowy and dataless. I fretted. On the sixth night we opened up and I reobserved the star, HL Tauri: the feature was real and quite as deep as the first set of data had indicated. One out of more than twenty young stars revealed ice. Why? It appears that ice is formed onto the dust grains in dark clouds where new generations of stars are born. As each star awakens and begins to heat up, it influences its environment to an ever-increasing extent, eventually melting the icy mantles off grains. This star must, therefore, be so extremely young that it has not yet succeeded in eradicating all the ice in its dusty cocoon. It may be only a few tens of thousands of years along in its evolution, making it one of the youngest stars known with mass comparable to that of our Sun.

There is a perhaps surprising aspect to living on top of a mountain range as far as the weather is concerned, namely that orographic factors (local phenomena associated with mountains) tend to make exceptional weather. Many nights it can be clear in the Arizona desert yet the high peaks are covered in cloud caps of their own making. It can be frustrating to realize that you could be observing if the telescope were situated a couple of thousand feet lower and a few miles away. Of course, when it is clear, a couple of thousand feet makes a big difference in the amount of water vapor sitting above your head. It is considerably drier at 9,200 feet on Mt. Lemmon than at 6,800 feet atop Kitt Peak, some fifty miles away.

Let us leave Mt. Lemmon, where we have been solely responsible for our own dining, welfare, engineering, and science, and visit Kitt Peak to examine the concept of the national observatory in operation.

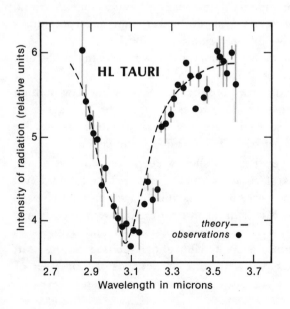

The near-infrared spectrum of HL Tauri obtained at Mt. Lemmon. The dots represent the observations and the vertical gray lines the errors associated with each point. That the absorption feature is due to pure water-ice grains of 0.3-micron radius is shown by the curve, which was calculated theoretically and fits the data very well.

KITT PEAK
NATIONAL
OBSERVATORY

6

Astronomy is big business these days. Universities commonly band together in droves to establish a major facility. Kitt Peak National Observatory (KPNO) is the product of such a consortium, AURA by name (Associated Universities for Research in Astronomy), not to be confused with AUI (Associated Universities Incorporated) which runs the National Radio Astronomy Observatory (Chapter 12), part of which is also set up on Kitt Peak.

The essence of the national observatory is that light pollution is woven so much into the fabric of urban and metropolitan life that most major universities have to build themselves observatories at some distance from their campuses and their support facilities. Why not set up one giant astronomical complex in a carefully chosen location, guaranteed to ensure good weather and clear dark skies, and maintained on site? So KPNO was born, in 1960, at 6,800 feet above the Arizona desert, in Papago Indian lands. The contributing universities are scattered from coast to coast, and the day of the backyard professional observer is gone. We have become jet-set commuters. In 1960 there were a few small telescopes on Kitt Peak, but in the past two decades KPNO has flourished to cover most of the flat, rolling, summit plateau.

We fly into Tucson, from our home base in the early afternoon, and take a taxi to the KPNO office and headquarters on the modern campus of the University of Arizona. The blanket of hot desert air claws at us as we flee the cooled cab for the air-conditioned opulence of the headquarters. In an hour or so we leave Tucson on one of the

Kitt Peak seen from the summit. Closest is the dome for the 158-inch; then, receding along the road, are the Steward 90- and 36-inch reflectors. Along the skyline ridge are, left to right, the KPNO 84-inch (with the 50-inch to its lower left), two 36-inch reflectors, and a 16-inch. In the distance is the volcanic plug of Baboquivari peak.

regularly scheduled carryalls that shuttle visiting astronomers, Tucson-based staff, and equipment, between the city and the observatory. The white-painted vehicle, sporting its proud navy blue, white, and gold insignia (a telescope dome surrounded by a constellation of stars, one for each member institution in AURA), smoothly races past beckoning saguaro cacti, spiky cholla, and ubiquitous creosote bush; past coyote; and toward the rugged green-brown skyline. We turn due south off the highway and attack Kitt Peak, on whose crinkled brow sits the massive white sentinel of the dome for the 158-inch telescope. Blue skies envelop us as the desert recedes, and mercifully the air cools a little. On the summit we thread our way through the crowds of tourists (KPNO is open 10 a.m. till 4 p.m. daily) to the mountain headquarters. We take deep breaths as we stagger with our suitcases to our motel-type dormitory room, our home on the peak for the next few days.

It is many months since first we conceived the present observing program. It had to be sketched out, pondered upon, written up coherently and convincingly, submitted far in advance of the six-month period in which we requested telescope time. The proposal was read by a committee of our peers, graded, and, if favorably reviewed, was allocated the time requested (at least some of it). The weeks slipped by, a time in which we prepared for every contingency and assembled charts to the class of objects under investigation. At last, as we join other astronomers in the mealtime trot to the diner, we have arrived! We do not start to observe until the morrow, but by overnighting on the peak we shall have many hours to prepare, in particular to check out the equipment that will be mounted for us on the telescope in the morning. Or, perhaps, we should like to look over the shoulder of tonight's astronomer if the same equipment will be available to us, to familiarize ourself with the instrumentation and to see if something can be learned from someone else's techniques.

The stream of people becomes a torrent as we queue for trays and silverware in the diner, after signing ourselves in for the relevant meal on a housekeeping sheet by the entrance. There are maybe sixty people here now, not all astronomers. It requires a huge supporting staff to maintain such a site: there are construction personnel, every form of urban maintenance person, office staff, a librarian, photographic specialists, staff astronomers, electronics and computer technicians, truck drivers, supervisors of the overall functioning of the community, police, people who run the museum and giftshop for tourists. Dinner is always filling, unless one is possessed of the willpower of an ascetic. The food is good and abundant. We seek some familiar face in the throng and sit down at a table with our selected person. Friendships are strengthened, acquaintances forged, introductions made. Chat is brisk: What are we doing here? Which telescope? What instrumentation? How many nights? When do we start? Do we know that so-and-so has been trying the same type of approach to a different class of celestial objects? Conversations are punctuated by twirling forks bearing gravied morsels; ice cubes chink in juice glasses; knives stab the air to emphasize a point.

It is always overwhelming to a newcomer, this crowd of celestial pundits, this active scientific harangue. It is often stimulating, and one can sometimes trace a subsequent idea to its vespertine origin at Kitt Peak. To meet other scientists is interesting, especially if their areas of research are different from your own. Even more fruitful is an encounter in which a common interest is discovered, but each person contributes a different technique, type of instrumentation, or spectral region. Discussion is generally lively, often even aggressive.

One has only to watch two rivals verbally sparring (hopefully only verbally) to appreciate that astronomers are human beings too, and that academia does not vouchsafe an escape from competitiveness. Far from this, science in North America is active, bouncy, professional, and aggressive, whereas it is always pursued in a more subdued manner in Europe. You need a quick mind, total alertness, much knowledge of rumor and grapevine gossip (scientific, of course) to hold your own in such a conversation, but you may find it invigorating. Next comes another demanding ritual: dessert. Kitt Peak proffers unlimited quantities of ice cream and all the necessities for its decoration. I find this irresistible.

Eventually we stagger away from dinner and betake ourselves to the well-kept little library in the main building to peruse the latest journals. Astronomical literature is exploding, a trend shown by all the sciences. Look at the shelves of scientific periodicals on any campus, and note how the number of shelf-inches per year has been increasing through the past decade. I envy people like Newton who were able to hold down all of physics, mathematics, and astronomy. In Newton's day so little was known about the world that one could acquire a working knowledge of all the physical sciences (and, of course, he was busily making these up as he went along.) Today it is a major task merely to keep abreast of who is doing what even in your own fairly narrow regions of interest.

Let us look at the journals, some monthly, some twice monthly. Most widely distributed in the world, and vital if one is to reach U. S. workers (whose reading is less catholic than that of European astronomers, perhaps for simple reasons of geography), is the prestigious *Astrophysical Journal*. Then, a basic list includes the *Astronomical Journal, Monthly Notices* of the Royal Astronomical Society (of Britain), *Astronomy and Astrophysics* (a European journal), *Publications* of the Astronomical Society of the Pacific, *Astrofizika* (English translations of a Russian journal), *Nature*, and *Astrophysical Letters*. Even if one were to digest only a single article in each journal, that would already be a time-consuming task. In reality one can keep somewhat abreast of the frontiers of astronomy by judicious use of the telephone to distant colleagues, and by attendance at occasional national astronomy gatherings. But, be assured, astronomical literature is ever-expanding, and you will need to know at least where to find important papers, even if you have not memorized their salient points and findings.

It is very dark to our unadapted eyes, an almost tangible darkness, as we grope our way from the main building toward the dormitory. But we have our image to preserve, so we stand a moment, quieting the noise in our eyes and brain; all astronomers can see in the dark. To the northeast the darkness glows, and Tucson sparkles across the desert in tangles of color. Tucson has grown rapidly, but so responsive has this city been to its local asset of astronomy (and, perhaps, for sound economic reasons) that recently cowls were installed over street lights to diminish the upward scattered light pollution. This is in sharp contrast to the sad deterioration of the San Jose sky below Lick Observatory (see Chapter 8). At 2 or 3 a.m., making an effort to switch our circadian rhythms to a schedule more conducive to observing, we finally collapse into bed with a good novel. There will be much to do tomorrow, and we will need our sleep tonight.

With adrenals operating, we rise for lunch, eat our way through salad, hot dogs, french fries, cake, fruit, juice, and find out whether last night was fair from some early-rising (or soon-departing) astronomer. We hurry away to our dome, refreshed and heartened by the blue clarity of the Arizona sky. There are several broad phases of our preparation: whatever instrumentation we are to use must be investigated *per se*, then checked again for possible remote operation under computer command, and finally the telescope and its connections to the piece of equipment should be examined. Computers, of some form, are being used with increasing frequency. At Kitt Peak, small computers are attached to all the telescopes of aperture exceeding thirty-six inches. These enable the telescopes to be offset from a star, of specified coordinates in the sky, to some nearby location. This procedure is vital for setting onto galaxies too faint to be seen by eye but of known position, and for daytime operation in the infrared. The computers, interfaced with instruments mounted on the telescopes, trigger numerous functions remotely. They store data and may even process them in real time, that is, while still being collected. In short, and this is the insidious aspect, it is generally impossible for a telescope or instrument to operate without an associated computer. Which means, in the event of computer failure, that everything may come to a halt. This is a sensitive and controversial issue in modern astronomy — how much reliance to place upon a computer and, relatedly, which is the most cost-effective and reliable small computer system? You will hear this topic debated agitatedly at mealtimes on the mountain, with anecdotal horror stories to spice the discussion. But enough of negativity; we are ready to observe once it is dark, and we permit ourselves the jubilation and optimism of the beginning of an observing run, mingled with hope and prayers for clear skies and flawless instrumentation.

Eating again; a quick ice cream dessert; a brisk walk to our room to don boots and whatever clothing we are ritualistically accustomed to wearing. En route we say good evening to a coatimundi. These creatures, overgrown cousins of the raccoon, consume astronomers' leftovers, and probably do not read the *Astrophysical Journal* (although they wear a puzzled and cynical expression as if they did). Venus is suspended against a pink sky as we enter the cool dome, bearing our finding charts and our enthusiasm. The night begins.

Let us now turn our attention to those long-suffering heroes of the peak, namely the telescopes. Each has a different character and is most frequently used with a particular type of instrumentation. The mountain abounds in telescopes, perhaps they even breed up there. The roster includes: the 158-inch ("Mayall" telescope), alias "the 4-meter," 90-inch (Steward Observatory of the University of Arizona rather than KPNO), 84-inch, 52-inch (endowed by McGraw-Hill for the use of the University of Michigan), 50-inch, two 36-inch and two 16-inch reflectors.

In addition there is the solar observatory. This alone is worth the price of admission, as some might say. It is a monster, brilliant white against the deep blue sky. Atop its vertical tower there are several "flats." These plane mirrors rotate to follow the sky so as to feed a

Coatimundi on Kitt Peak.

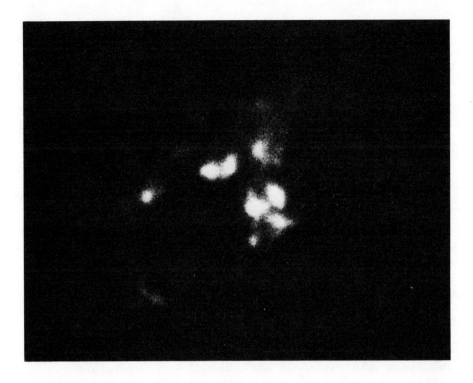

Herbig-Haro object No. 2 in Orion as photographed in red light by George Herbig of Lick Observatory, using the 120-inch reflector. Objects of this type consist of one or several semistellar nebular patches. Lick Observatory photograph, copyright University of California.

beam of radiation down the long inclined tunnel to the fixed pieces of giant equipment far below ground level. One of these flats has also been used for infrared planetary work during daylight hours. There is a viewpoint, halfway along the inclined tunnel, through which visitors can peer inside the huge cavern, up to the flats, down towards the business end. I recommend the view.

I have worked mostly on the 50-inch, a reflector modified and optimized for infrared work. It has long had computer control and this has been extremely valuable to infrared astronomers. During my latest trip to Kitt Peak, in late 1978, I and a colleague relied heavily upon this automation in pointing. We wished to search for infrared sources in the vicinities of several curious nebulous patches in the sky. These "Herbig-Haro objects" (named after the joint discoverers of the first ones in the early 1950's) have been surmised to represent either a reflection of a young T Tauri star off a dusty cloud or the result of shock waves raised locally when a stellar wind (like the solar wind but a million times more energetic) from a young star impacts upon its dense cloudy environment. One way to distinguish between the two hypotheses may be through infrared observations. These objects, like their putative allied young stars, frequent dark obscuring clouds that appear as vast empty patches on photographs made even with the biggest telescopes. However, from the deepest plates taken with large instruments it is possible to determine the right ascensions and declinations of these bright Herbig-Haro nebulae (by reference to bright stars of known position on the same plates, albeit most of a degree distant on the sky). With the 50-inch we cannot see these faint fuzzes, but we are able to point the telescope at the visible stars whose coordinates we know. Then we merely tell the computer these coordinates when one such visible star is set up in the infrared beam, and request it to drive the telescope to the coordinates of the Herbig-Haro object. If we find no infrared source precisely at the location of the nebula, we can instruct the computer to raster the telescope in a predetermined fashion, pausing at each of a number of locations to integrate in case a source is found there. If we do find one, we can narrow down its position by changing the infrared beam to a smaller and smaller size and making a very localized raster.

The 50-inch dome has an entire floor that rises and falls, which conveniently alleviates much of the usually required acrobatics on ladders. The dome, however, is possessed of one annoying feature, exacerbated by the offset telescope mounting: it has an extremely narrow slit, little wider than the primary mirror, and it is frequently difficult to tell whether one is observing a piece of the dome or not. In the infrared, where the dome is a strong emitter, you must continuously monitor the relative positions of dome and telescope. At Mt. Lemmon (where the dome is superwide anyway, about seven feet) this problem is solved by a ring of six red-beamed lasers, attached to the top of the telescope tube and shining towards the dome. (Actually, the 50-inch does boast a couple of anemic red lamps for monitoring the dome.)

By late 1978 a television camera had further been added to the infrared photometer to provide remote acquisition of objects. It is a delightfully easy procedure to locate a bright star on the television, tell the computer its coordinates, and ask for an offset to some invisible nebula, where one can automatically scan a small box around the object to see if infrared sources are hidden nearby. One felt in

December, 1978, that astronomy was catching up with technology! (And, incidentally, the computer now keeps track of and adjusts the position of the dome too.)

Much of my early work, carried out on Mt. Lemmon in the ten micron region, was devoted to broadband measurements of young stars. Recently, infrared detectors have improved, technologically. Now one can reobserve these same stars, but subdivide their overall ten micron radiation into forty or fifty small bandpasses, and detect each small piece almost as easily as the entire ten micron flux only five or six years ago. However, to insure that a sufficient number of young stars can be observed, a telescope greater than sixty inches in aperture is preferred. I have, therefore, carried out some work on the KPNO 84-inch. To put matters bluntly, the weather has been grotesquely unkind.

Most of my Christmases and New Year's Days since 1970 have been "celebrated" atop Mt. Lemmon or Kitt Peak (principally because my family has always been 6,000 miles away). I have now formally renounced this December habit, and it is my earnest intention to be far from Arizona (and its attendant rain, sleet, cloud, snow) during future holidays. But, masochist that I am, I tried this program yet again on the 84-inch for the third and hopefully final time in early December, 1978.

I ought not to mention the 84-inch without talking about its autoguider. The issue of guiding is a crucial one in the realms of sequential, as opposed to multiplex, spectroscopy. Let me explain some terms. First investigations of a class of objects are nearly always via photographic plates, which are broad bandpass detectors that receive all wavelengths across a large chunk of the spectrum simultaneously; that is, in a "multiplex" fashion. Next, one may obtain multicolor photometry, for example ultraviolet (U), blue (B), and visual (V), by placing filters over some detectors that define narrower bandpasses than do typical photographic plates, although still we are receptive to perhaps a thousand angstroms of spectrum. Further down the line, at least for infrared work, comes the "CVF" — a continuously variable filter.

The essence of these spectrometers is a cunningly designed filter that is continuously variable; in other words, it provides a spectrally narrow filter, whose wavelength varies from two to four, or from eight to about fourteen, microns, as the filter wheel is rotated. The procedure is to observe a single wavelength (one position of the wheel) at a time, then to step the wheel a small amount and observe the next wavelength. One observes a standard star whose intrinsic energy distribution is known, then a young star, and divides the two spectra, wavelength by wavelength. This technique removes the terrestrial absorption features (really only a single major one around ten microns, due to ozone) and is identical with the mode of observation that revealed the ice band in HL Tauri (Chapter 5). **UBV** data are sampled one filter at a time. Such processes are sequential, and if there are cloudy conditions they will affect each filter, and each spectral data point, to an unpredictably different degree. Furthermore, if one guides on the object erratically, with respect to some crosshairs that delineate the spatial center of the detector's response, each point in a filter-wheel spectrum will receive a different amount of stellar radiation. There do exist multiplex infrared spectrometers but, for early reconnaissances, filter wheels offer the advantage of permitting an appropriate amount of time to be devoted to each small piece of the spectrum. Consequently, consistent guiding (and clear stable conditions) is of the essence in filter-wheel work.

December of 1977 saw me back on Kitt Peak, gaining weight, playing pool, and waiting to open the dome of the 84-inch. I had been allocated five nights to use a ten micron filter-wheel spectrometer on my beloved young stars. I actually did observe, albeit only on the equivalent of one night. I think the elements were then tiring of me; after all, they had just cost me all seven nights of a 60-inch infrared run at Mt. Lemmon. For the reasons outlined above, I had been worried about guiding on my faint, isolated, young stars while operating the filter wheel. You see, young stars are gregarious only on the large scale; take a nearby nursery like the Taurus dark clouds, and one often finds no star brighter than thirteenth magnitude within half a degree of a T Tauri star. I knew I could not guide on-axis with the spectrometer unless I had a visually quite bright star, so I would need to use an offset guider. That device would permit me to set a fiducial mark on some star outside the infrared field of reception, to represent the centering of the star actually being observed in the infrared. Since the 84-inch had its top end modified for infrared, the focal ratio was very great, around f/26. Thus, the maximum physical distance that the offset guide micrometers could move was contracted with this reluctantly convergent beam to only little more than a few arc minutes in either direction.

There were times when I cursed the guider for its small field, times when I simply could not work on some isolated young objects. However, when there is an available guide star, if only a faint one, the guider performs valiantly and, hopefully, consistently, treating the guiding identically at all wavelengths. The autoguider is an optical device that senses the direction of motion of a visible offset guide star and emits drive pulses to the telescope, so as always to return the image to some previously defined field center. This was my first experience with an autoguider, and, as an amateur astronomer/masochist — you know the type: someone who really enjoys freezing to death while pushing eye and brain to the limit — I was always suspicious of anything other than my own eye for guiding purposes (even somebody else's eye). But, the autoguider works beautifully. As the atmosphere kicks the faint stellar images hither and thither, the two-directional drive to the telescope clicks in response, keeping the stars centered and their infrared energy falling squarely onto the filter-wheel and the detector. It is at first slightly discomforting not to be able to see your program star at all times, but the autoguider permits the observer to keep abreast of the raw data as they arrive via the computer terminal, wavelength by wavelength, making it easy to ascertain from the point-to-point smoothness of the raw spectrum that the star is accurately centered. Wonderful inventions, autoguiders, even if I have to grant that such an admission causes my amateur component to be dragged, kicking and squealing, into the 20th century.

The results of the ten-micron study of young stars were also pleasing. About ten years ago I tried to model the infrared energy distributions of T Tauri stars and related young objects by assuming that circumstellar silicate dust grains were present. It was a plausible guess, but only a guess. Since then I have been very critical of people who attempted to identify these materials in young stars on the basis of the most tenuously suggestive data. The observations were obtained with filters all too broad in bandpass and with inadequate signal-to-noise ratios, and the data were interpreted without troubling to demonstrate that the possible weak features in young stars resembled those strong features generally attributed to silicates in other classes of stars. However, I am now a comfortable believer. By early 1979, after careful analysis of the 84-inch spectra, and after comparison with the acknowledged silicate spectra, I could claim that T Tauri stars showed silicates, sometimes in emission and sometimes in absorption. My conclusion was a long time coming, but I'm glad

The dome of the 158-inch (4-meter) telescope.

that I persisted during those three miserable Arizona Decembers.

On the outside it resembles a set from the opening of Woody Allen's *Sleeper*. It is huge, latticed by girders, and brilliant white against the deep blue sky. It houses Kitt Peak's 4-meter telescope (158 inches but reduced to 150 inches for direct photography, since a few inches of the rim of the primary mirror are turned down). The reason for the extreme height of the structure is to avoid low-level turbulence, as air flows over the rock pile that represents the summit of Kitt Peak (good scrambling from the east), or over the edges of the mountain plateau, or past Steward Observatory's 90-inch (so small and so far below when viewed from the catwalk of the KPNO 4-meter). As one drives around the final rocky lump on the road up to Kitt Peak, one passes some massive bolts, connected to steel plates. These were essential in order to provide a solid enough foundation on which to base the telescope. You can easily get lost inside the dome. There are two elevators, countless rooms, corridors, several floors, workshops, darkrooms, lounges, a pool table, a kitchen, a library. When the next biblical-style flood arrives, head for the dome of the 4-meter; it could support a small community.

The telescope, too, is a beautiful sight: short, stubby (it has a fast f/2.5 primary), and painted bright orange and white. Its Cassegrain cage is big, though not as roomy as Palomar's (see Chapter 10). It is run remotely from a computer-equipped control room, thermally isolated from the observing floor, and an acquisition TV is provided. But the fanciest aspect of the telescope is its "flip ring" top end. You see, to provide separate top ends for the prime-focus cage and for the Cassegrain, infrared-optimized Cassegrain, and coudé secondaries is very expensive and involves a specific design for the tube. Secondly, it has always taken a matter of hours to change top ends at observatories such as Lick and Palomar. However, there are often times when the seeing or transparency changes radically during the course of the night, and it could be more efficient to use a different focus than the one for which the telescope had been set up at the start. The 4-meter is ingeniously configured. Stow it horizontally; connect a floor-mounted drive shaft to a hole in the topmost ring of the tube, and you can rotate an inner ring through 180 degrees. As this ring flips around, it interchanges prime-focus cage and Cassegrain secondary in fifteen minutes: fast, efficient, and much more economical than building separate top ends and an attendant overhead gantry crane.

If the telescope is a gem, then just as surely its scientific output has been compromised by computer failures and problems with auxiliary equipment. The crucial issue is far from hypothetical: you run Kitt Peak; your budget is fixed at about $8,000,000 this year; how do you allocate the money among so many different telescopes, ancillary equipment, and possibly fruitful new programs of instrument development? Not an easy set of decisions, perhaps explaining why it has recently required two years to elect a (willing) new director for KPNO. But as a research weapon, Kitt Peak is very important. If one is a teacher at a small school or college, KPNO is all-important, a place to escape to for some time in the summer and a place that provides telescopes and weather for all institutions with no access to observatories of their own.

A guaranteed topic of discussion at any observatory is the question of the seeing. At midnight lunch on Kitt Peak one can find perhaps half a dozen observers with different telescopes trading horror stories. One may claim his ninth-magnitude standard star swelled up to an

The KPNO 158-inch parked horizontally.

image size so gruesome that you couldn't even see it through the eyepiece. Another, working the coudé focus of one of the large instruments, will know when the seeing is very good because then all the light from his star will fall into the slit of the spectrometer, leaving none to be reflected back to the guiding eyepiece off the polished slit jaws. These days, with increasing dependence upon television cameras for remote operation, acquisition, and guiding, it is harder to obtain estimates of seeing than in the days of eyeball guiding. Simply by turning up the gain on a television camera one can produce a bigger image, so seeing is no longer a uniquely determined parameter. Furthermore, if you have to observe a faint galaxy, then you will use a long integration period in the television memory. This tends to average over a longer interval than does the combination of eye and brain and, hence, reduces the apparent motion of the image.

Another problem is that seeing may vary from place to place even on the same mountain site. I have often heard arguments at Kitt Peak between, say, observers on the 50-inch (in the middle of a flat plateau) and the 84-inch (at the edge of an abrupt cliff) during which claims were made for seeing of quite different characters. There is a simple explanation to this paradox. At least one component, namely the variations of image size as opposed to the side-to-side motions of the image which can be quite independent of the size, seems to be widely attributed to atmospheric cells very close to the ground. The motions of these cells clearly are influenced by the nature of the air flow in the immediate vicinity of the dome. One might, therefore, expect smooth laminar air flow to produce good seeing while turbulent flow would yield poor image quality. Such a model provides an understanding of the relationship between seeing quality and the specific location of a dome at an observing site.

In fact, I well remember one evening, early in the history of the Mt. Lemmon Infrared Observatory, when a vehicle drove up to the dormitory and two astronomers alighted. Would it be all right if they launched a balloon from our site? I responded affirmatively and asked what was the purpose of this balloon flight. They wished to carry aloft a smoke canister and to release it over our part of the mountain. I had momentary visions of dark serpentine pillars of smoke writhing biblically over our dome. Their experiment was actually related to a determination of the air flow on Mt. Lemmon, with a view to determining the best locations for future telescopes. The convolutions of the smoke were recorded with a movie camera and revealed a beautiful laminar flow from the summit down to our infrared telescope, a hundred feet below. However, at the location of at least one other telescope, closer to the summit and to a steep drop that commonly faced the prevailing wind, there was evidence of much turbulence. Indeed, users of that telescope often complained of poor seeing on nights when we enjoyed modest or good conditions.

At this point, I should like to describe a fine research instrument to you, one which is to be found on Kitt Peak but which is taken very much for granted and used solely as a preparation for other observing. I am referring to the fine set of deep photographs that constitutes the National Geographic Society-Palomar Observatory *Sky Survey*, acquired with the 48-inch Palomar Schmidt telescope. It was December of 1976, just after Christmas, and Kitt Peak was "socked in." You couldn't see a dome, let alone a star. It was cold; the wind was scouring the peak horizontally; a few snowflakes were wandering about uncertainly. It was warm, comfortable, and light in the *Sky Survey* room in the dome of the 84-inch. I cleaned my 10x eyepiece, with its calibrated reticle, and prepared to do something I had wanted to do for years.

As a graduate student in England, part of my dissertation work was to identify photographically infrared sources on the *Sky Survey* photographs. My English advisor was really *au fait* with the photographic appearance of a host of different objects. I profited much from his experience. Of course, for several months I did spend up to six hours a day, five or six days a week, poring over the photographs. I could tell a photographic blot from a planetary nebula (both are circular and have abrupt edges); I could differentiate real from spurious obscuration (the Eberhard effect, a local starvation of developer in the vicinity of bright objects that yields a whiter-than-white region); I knew a faint irregular galaxy from a thumbprint.

I, along with a Berkeley friend and colleague, had sunk some four years of infrared and optical effort into a multifaceted attack on star formation. Our desire was to obtain low-resolution, quantitative, optical spectra with the Lick Observatory image-tube scanner and also broadband infrared data on a very large sample of young stars. The optical spectra would yield effective temperatures of the stars, on the basis of absorption features produced in the cool outer layers of the stars — their photospheres; the infrared would yield reasonable estimates of their total luminosity, by extending the tiny piece of visual energy to far longer wavelengths (much of the luminosity of a young star can occur in the infrared). These two crucial stellar parameters characterize a star and enable direct comparison between theory and observation. But where would you start?

First, you seek previous surveys of the sky for stars that are believed to be young. In youth, stars exhibit bright emission lines of hydrogen, the strongest being the red H-alpha line of the Balmer series. Then, if you photograph a portion of the sky through an H-alpha filter and compare it with another photograph taken through a broadband red filter, it is straightforward to identify objects that seem curiously bright in H-alpha. Unfortunately, not every object that shines brightly in H-alpha is a young star; for example, planetary nebulae are also bright hydrogen emitters. However, if one has surveyed a region of thick dark clouds, it is unlikely that the line of sight can penetrate through the obscuration. Therefore, the bright hydrogen objects either lie in front of the clouds, which is also unlikely if the clouds are close to us and there are many hydrogen objects, or inside them. If they lie within the clouds, then these objects have an excellent chance of representing recently formed stars still involved in their dusty nurseries.

Cryostats at Kitt Peak National Observatory. The boxes attached to their sides accommodate preamplifiers; other appendages are for changing the aperture or filter or are vacuum ports.

Therefore, we began by observing all stars thought to be young that had been revealed during previous hydrogen-emission surveys. That gave us about 300 stars. During observing trips to Lick Observatory, I would make a habit of looking at the *Sky Survey* photographs in the daytime, seeking hitherto unknown, faint, red stars in the dark stellar nurseries. With a combination of intuition and trained eyeball, I could always find a couple of young stars purely by photographic inspection. The problem intrigued me. How many unknown young stars were there, lurking in clouds and awaiting recognition?

I had six nights of 84-inch time allocated to ten-micron spectra in Decembers of 1976 and 1977, and the storms raged throughout. Each day, however, I "observed" on the Schmidt photographs. There are 879 pairs (red and blue) of *Sky Survey* plates, and an additional 100 (red only) survey plates at southern declinations below −30 degrees. I looked at them all. My desire was to cover completely all dark clouds, and, if possible, to peruse each photograph in its entirety. Now, for those of you who have not seen a Schmidt picture of a typical field in the Milky Way, let me enlighten you. Take a fourteen inch square piece of pale grey paper; sprinkle it liberally with a couple of pounds of fine black sand (try the volcanic beaches of Hawaii); deposit a few globs of thick, white paint randomly over the paper. This is a fair representation of the Milky Way. It is absolutely dense with information. In fact, it is too dense for perusal. In such cases, I concentrated on the dark-cloud areas (the globs of white paint — the *Sky Survey* is issued as negative prints, with black stars on a white background.) I noted some 200 red, nebulous, or otherwise curious objects. Then I had to remove from my list all objects known from other catalogues, only a slight reduction. I had my answer at last. Slowly, during optical trips to Lick (Chapter 8), we began to take spectra of my "dark-cloud objects" and turned up a significant number of unknown young stars, some new Herbig-Haro objects, and a large number of hotter stars that revealed, at most, weak hydrogen emission.

By far the most abundant stars in nurseries are stars with a mass less than or equal to that of our sun. So where are hotter, more massive stars born? (These are the stars of spectral types *O, B, A, F* once they reach the main sequence.) I am still trying to unravel this issue, but the high percentage of these stars in the dark-cloud survey may indicate that there are many, small, distant clouds, in which stars of all masses are produced but which reveal only the most luminous ones (those of higher mass and hotter temperature) when observed from afar. It could be that we obtain a highly biased picture of star formation by concentrating our efforts on the nearest nurseries, the clouds of Taurus and Auriga (500 light-years distant), which have recently produced so many very young stars of low mass. For the desperately curious reader I will say that there are many places in which our galaxy has produced hot stars in abundance; and perhaps low-mass stars too, although these often would be too faint to have been observed yet at the relevant distance. Nevertheless, a crucial question, as yet unanswered satisfactorily, is: does each nursery yield stars of all masses in the same proportions? If so, how? If not, why not?

So ended my first two runs on the 84-inch. At least the time was used constructively! Of course, not every period of bad weather has sent me scurrying, eyepiece in hand, to the Schmidt photographs. I have played my share of pool on Kitt Peak, and bridge, and hearts, and "guess the author" (played with titles of articles in astronomical journals). And then, there is the monsoon.

If one is to become an observer, one should be cognizant of the Arizona monsoon. Even cacti need some water, and the rains of the

Arizona-Sonora desert clime are concentrated into brief devastating periods, principally in the summer months of July, August, and September. Each afternoon, thunderheads build up around the mountains, driven by the massive thermals boiling off the vast hot sands. Each night the sky is alive with flashing, and only in the hours before dusk does this activity diminish, to be replaced by often clear daybreaks. The clear skies offer but false solace, however, and by midafternoon the pattern repeats. Domes must be protected against lightning, telescopes covered with waterproof bibs in case of leaky domes, computers shut down totally. Monsoon is a highly frustrating period for astronomers. One year it may thunder and rain in this fashion continuously for three weeks before permitting a brief period of clear weather. The next year may bring largely clear skies with only a few sporadic flash floods and storms.

The monsoon also bears natural conditions that are honestly termed spectacular. I recall several such phenomena. South of Kitt Peak the great thumb of Baboquivari thrusts skyward, the exposed solidifed core of an ancient volcano. From above the desert it is possible to perceive that rain really advances in sheets — ominous, gray, twisting sheets. Baboquivari with its knob wreathed in folds of rain is dramatic. To be standing on the outer skin atop the dome of the 158-inch, with lightning flickering, orange, sullen, and silent on three horizons is another unforgettable product of summer observing. Perhaps the most beautiful sight I have seen on Kitt Peak — excluding the raw data spectrum of HL Tauri at ten microns, revealing a silicate absorption feature to go hand in hand with the three-micron water-ice fingerprint! — is a rainbow.

I had been working with one of the graduate students from Berkeley on the KPNO 50-inch, scouring dark clouds in Cygnus for embedded infrared sources and making computer offsets from visible stars to the locations of optically faint hydrogen-emission stars, all a part of the magnum opus on star formation. It was a very frustrating trip, for we would only have short clear spells between total cloud, showers, and lightning. One night we actually worked for no less than five hours in quite clear conditions! I shut down the dome close to sunrise and went to bed. I awoke after half an hour's doze to hear a gentle tapping at my window. I staggered out of bed, elevated the black-out blind, and beheld a wondrous sight. The sun had just risen; the mountain's brow was sprinkled with wispy clouds, painted orange and brown by the low sun angle; large droplets of rain were descending,

Double rainbow at sunrise.

and a rainbow was clearly in the making. I seized a camera and a wide-angle lens and went outside to watch. From my elevation I could see more than the usual rainbow — I could see the bow bending back on itself against the distant valley far below me, revealing more than a semicircle. The sky was pink and purple, and both primary and secondary bows were strongly colored. I twisted my polarizing filter around, noting the strong, radial polarization of the double arc of light. It was a thrilling performance and you can share a little of it with me in the picture above. Incidentally, if you are ever observing with the No. 2 36-inch telescope, you might take a pick and shovel and see if there is a pot of gold below the floor of its dome . . . .

The evening is very still, there is no wind, the sunset colors linger in the west. The young night is crossed by the trundling sounds of opening domes. Dull reddish glows leak from half-open slits and a jeep crunches noisily up the hill to the 84-inch. Overhead arches the crisp Arizona sky, dark blue, clear, star-pocked. Coatimundis rustle through the shrubs by our dormitory window. We collect our charts, don boots and down jackets, and stamp out into the air with elation and excitement. It's going to be a clear night on Kitt Peak, and there's a whole universe of mysteries waiting for us.

LEUSCHNER
OBSERVATORY

7

San Franciso Bay is buried under a white shawl; it's up to its high-rises in fog. The winter weather pattern is established. We drive over the misty ridge of the Berkeley hills, heading east into the countersunset, for Leuschner Observatory. A fast focal ratio 30-inch reflector is housed in a small dome about twelve miles over the hills from the University of California campus at Berkeley. Principally a teaching instrument, the 30-inch is, nevertheless, capable of useful research. It is January, 1975, and I have let a colleague inveigle me into slaving one night over a cold **UBV** photometer to define a three-color light curve for Eros, the tiny asteroid tumbling through space enroute to opposition on the 13th. Initial preparations were brisk and straightforward, consisting mainly of flailing a wooden mallet at a sack filled with dry ice to pulverize the chunks for feeding into the photometer's cold box.

We had a photocopy of the *Sky and Telescope* chart on which was plotted the path of the asteroid. Preliminarily, several standard stars were observed for calibration. We located Eros easily. There then ensued a seven-hour marathon of ultraviolet, blue, and visual measurements of Eros, repeating the filter cycle every few minutes. The only "breaks" we took were to observe standard stars each couple of hours. The end products, reduced by means of a ruler to measure deflections directly from the strip-chart recorder and fed to a large computer for neat and rapid analysis, were three beautiful light curves. I recall that I derived as much satisfaction from those curves as from any set of ten-micron observations of strange variable stars! Quite

The fast-focal-ratio 30-inch reflector at Leuschner Observatory, equipped with an infrared Michelson spectrometer at the Cassegrain focus. Note the rubber hose, which maintains a very low pressure, hence low temperature, on the liquid helium in the cryostat.

a chunk of rock, Eros, and clearly pointed at one end and snub-nosed at the other as evident from the two very different minima — one abrupt and short-lived, the other gradual.

Another visit to Leuschner that is indelibly engraved on my mind (as well as on the soles of my feet) relates to Comet 1973f, alias Kohoutek. It was our hope, at Berkeley, to take the infrared interferometer (described in greater detail in Chapter 13) and observe the ten-micron region of the cometary spectrum. At this stage, two types of cometary emission were known in the infrared: featureless spectra, surmised to come from iron particles, and spectra revealing a signature at ten microns, identified with silicate grains. We believed Comet 1973f had silicates and hoped to narrow down the identification. It was a frustrating day. We began observing at about ten o'clock in the morning (there is no infrared night, remember) and tried to set the telescope beam onto the comet. The only manifestation of success would have been a deflection on a chart recorder, responding to the flux in the entire ten-micron window through our atmosphere, prior to feeding the beam into the interferometer. Every hour or so we telephoned Minnesota, where their 30-inch was busily observing the comet with a broadband infrared photometer, and whose coordinates we used, in preference to the ever-difficult interpolation of an initially estimated ephemeris. We hunted high; we hunted low; we pointed the telescope with intended precision; we shook it in rage; we criss-crossed the expected path — nothing. As the sun set that evening and the sky darkened, I found the comet, using my naked eye and later the small finder, but by that time we were at too extreme an hour angle to afford any time for infrared spectroscopy (and our feet were numb with cold). Still, the Leuschner 30-inch is a handy test-bed for new equipment and, given enthusiasm and perseverance, is a good research weapon.

8 LICK OBSERVATORY

Squirrels sprint across our path as the road serpentines through woods dense with live oak and manzanita. Below us the smoggy bustle of San Jose falls away and there is only the road, the trees, and blue sky. Raw materials for the original Lick Observatory were carried up to 4,200-foot Mt. Hamilton by horse power, and the tortuous highway with its gentle gradients is a legacy of that era. We traverse a ridge and plunge into a shadowy valley spanned by a broad blue lake. Ahead of us, on the skyline, sparkles a row of bright white domes. This first glimpse is deceptive and some tens of minutes have elapsed by the time we crest the final ridge, below the patronizing stare of the fine old building that once housed the Lick astronomical community.

We park outside the little dining room, collect our dormitory and dome keys from the kitchen where the odors of lunch induce instant salivation, and transfer belongings to our rooms. Lunch is a tasty intimate proceeding (since some of last night's observers may still be abed) with a handful of observers and perhaps some of the resident engineering personnel. There are several telescopes in frequent use: the giant 120-inch "Shane" reflector built in 1955; the historic 36-inch refractor (the original Lick instrument); the famous 36-inch Crossley reflector (imported from England at the turn of the century); a 20-inch astrograph (used to provide a mass of plates for determination of proper motions and now into its second epoch program); a 24-inch reflector (recently instrumented in a remarkably state-of-the-art manner, although this equipment will be transferred to the just-finished 40-inch). In addition, there are two smaller, less frequently used in-

The Lick 120-inch reflector in its massive fork mounting.

struments, a 12-inch refractor and a 24-inch reflector. The former is historic and has been ousted from its ancient wooden dome to make room for the new fast 40-inch; the latter is a prime telescope for sky-gazing. There used to be a flourishing community of perhaps a hundred souls living atop Mt. Hamilton, including a fire station and a school for twenty or more young children, until in 1967 the University of California opened a new campus at coastal Santa Cruz, some forty miles from Lick. At that juncture the astronomers and their families left the mountain; astronomers and library moved to the fledgling campus to become the astronomy department, leaving a skeleton staff of engineers and maintenance personnel and an echoing building.

After lunch, we drive to the telescope that we are scheduled to use tonight and set up our instrumentation. When I first began to use the Lick telescopes in 1972, it was to do long-wavelength (ten-micron) infrared spectroscopy and photometry on the 120-inch reflector. Mt. Hamilton is a low-elevation site for this spectral region, and subsequently I have come to know the 120-inch as a spectacular optical spectrometer. Indeed, I have used Lick as the optical counterbalance to my infrared studies from Arizona in the long investigation of star formation. Let us visit with this spectrometer.

The dome is large and quite cool, even in the summer. The telescope is massive, currently yellow-painted, and it swings in an incredibly large fork mounting. Around the dome floor lie several different "top ends" for the openwork tube: there is a prime-focus cage for direct photography, but it is little used now and looks every bit like some archaic torture device; a Cassegrain secondary; a coudé secondary and coudé flat, replete with complicated gearing to pipe a beam through a fixed hole out of the dome to the giant concrete coude room below. We shall be using the Cassegrain focus. We drive around the dome and attend the backplate of the telescope, using a powerful fork-lift truck nicknamed "the cherrypicker." The giant spectrograph is already mounted for us, looking every inch the canonical black box; it is, in fact, a black tetrahedron, supported from its base and hanging point downward, and accompanied by the churning sounds of a small electric motor that assures us that the device is being cooled.

Why is this black box so remarkable? The heart of the device is a chain of three highly sensitive image tubes operating in sequence. An image tube consists of a photocathode — a chemical layer at high

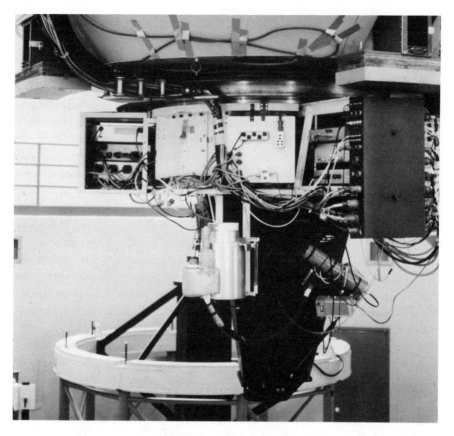

At right, sticking out of the black box at an angle, is the image-tube scanner, attached to the 120-inch telescope's Cassegrain focus. The fat aluminum can in front of the spectrograph contains coolant that is circulated around the image tubes. On the floor of the dome, beyond the telescope, is another top end for the 120-inch.

voltage with the property that when photons of radiation impinge upon it electrons are ejected. Each of these electrons is focused to the other end of the tube where it is incident on a phosphor and yields many photons (like a TV screen), hence the gain of the device over conventional light detectors. A multistage tube simply connects several such devices in series so that the phosphor at the back of one tube is pressed against the photocathode at the front of its neighbor. This produces a multiplication of the gains of each of the tubes, and a total gain in sensitivity to light typically of tens of thousands. Often a photographic emulsion is pressed up against the final output phosphor, but this is not an efficient process. The Lick system adds another stage that rapidly electronically "reads" the decaying glow on the final phosphor and feeds this scan into a computer where the information is stored. As time goes by, each rapid electronic reading of the final phosphor image is added to the computer memory, producing a means of storage (integration) for long periods.

There are several highly convenient features of the overall system. The dome is kept totally dark; observers and telescope operator work in a warm, lighted room, isolated from the dome. Guiding is by means of a television camera mounted inside the telescope and displaying an image of the sky field on a monitor TV screen. Two pieces of sky, quite close together, are fed simultaneously to the spectrograph in the bowels of the black box, and the resulting two spectra are both electronically scanned and stored. This is analogous to the infrared chopping technique (Chapters 4 and 5), for one patch of sky includes the star and the other does not. The mode of observation is to switch the star, after some minutes, to the other slit of the spectrograph and integrate there for a while. The computer generates the difference between the two spectra, and this difference can be augmented every time the star switches beams. In this manner it is possible to see, while the integration is actually under way, whether adequate data have been recorded or whether a longer integration would be profitable. The skies are bright over San Jose, and they steadily grow brighter as the city increases its sprawl across the valley each year. Subtraction of one spectrum from another is essential to cope with this bright background; the night sky also includes a spectrum of emission lines due to street lights with mercury and sodium vapors. Another important aspect of the system is a real-time display of the two spectra. Suppose you are searching for a particular type of star in a crowded field, as if you wished to carry out a survey of a cluster for one special stellar characteristic. This real-time spectrum can rapidly show which stars are of the spectral type desired and merit observation, and which are not relevant to the program.

It is early afternoon. We are in the cool dome, riding the cherrypicker, filling a flask with a mixture of alcohol and dry ice (frozen carbon dioxide) that is circulated around the image tube chain to provide cooling and a substantial reduction in electronic noise. We check inside the mysterious black box that houses the spectrograph to insure that the correct diffraction gratings are installed for our purposes, depending on the dispersion of the spectra we want and the total wavelength coverage. It is time to test and calibrate the instrument in preparation for the night. The dark slides are opened, and we return to the control room to titillate the computer. The principal calibration is that of wavelength. Information on the spectra will be referred to 2,048 channels, or different positions, on the phosphor of the image tubes. To what wavelength does each channel correspond? Inside the spectrograph housing there are lamps that emit line spectra at precisely known wavelengths (helium, argon, neon vapors, typically). By recording such lamp spectra before we begin to observe, we will determine the precise wavelengths of several channels across the image tube. The rest can be determined by interpolation — alas, not linear! Routine calibrations should not exceed one or two hours. Then it is time to plan the science.

Time on big telescopes is only obtained by a competitive process, and it behooves us not to waste a precious moment. We carefully sketch out a program and list the positions of the first few stars for the use of the telescope operator, who actually sets the instrument in the vicinity of each object for us. We then locate the star precisely on the TV, with reference to its neighbors on our star charts, and we are ready to take its spectrum. After an estimate can be made of exactly how much time needs to be spent on the star to obtain adequate signal-to-noise in the spectrum, the telescope operator is told what will be the next object on the program. The lights flash; the computer indicates it has completed the acquisition of the last few minutes of spectrum; the TV is shut down; the telescope slews toward the next star; the person at the computer adds the ultimate spectrum to the accumulated previous ones and displays on a screen the total information gathered on the object. While a paper copy of this spectrum is being plotted by the computer, the TV is back in operation, scanning for the next star. This procedure, I feel, should be precisely as crisp as described above, and this efficiency should be maintained throughout the night. It is not easy but it is very reward-ing, and there comes a point at which everyone and everything has been blended with the telescope into a single-minded machine: ob-servers, telescope operator, computer, spectrograph.

It should be noted, however, that instrumental problems occasional-ly arise, and several hours of a clear night can be jeopardized, even totally wasted. The spectrograph and image tube are extremely com-plex mechanisms, and many independent electronic units must meld together flawlessly to insure continued operation. After twenty or

The 120-inch control room with the author seated at the PDP 8/I computer, which with its peripherals occupies the right three electronics racks. In the foreground is a TV screen used to acquire and guide on stars.

more nights on the instrument, one becomes able to recognize a small class of common problems that can be readily diagnosed and cured without too much difficulty. Of course, one is always somewhat loathe to turn on the dome lights to effect repairs, for even with several forms of shutter and dark slide the image tubes can always sense the background of light. Even after dark has been reestablished, the tube phosphors will be noisy as the glow slowly decays. The most stomach-curdling computer message involves the flashing on and off of every lamp on the control board for computer and spectrograph, to the accompaniment of a high-pitched "beep" from the input/output terminal, while the display screen flashes the word "DISASTER" in green letters two inches high. Not a pretty sight.

At the end of the night, the TV is turned off, the telescope is stowed pointing at the zenith, the dome is closed, and the calibration procedures for the spectrograph are repeated. It is then bed time.

The Crossley dome is separated by about a half mile from the other telescopes at Lick. We open the door creakily, and a smell of mustiness assails our nostrils. The Crossley tubing is old, and it looks it. With its riveted iron plates, it could have come from something out of Jules Verne, as recreated by Disney. Almost everything else about the instrument is of much more recent vintage, completing a long drawn-out process of upgrading the telescope. The Crossley is "historic," but all that actually remains of the original bequest of many decades ago is the concept, to quote the present Director of the observatory. The mirror was replaced many years ago, as were the tube, finders, drives — nothing is original. If the 120-inch is operated remotely, antiseptically, then just as surely the Crossley is not. If you are feeling isolated from real telescopy, apply for Crossley time. There is nothing antiseptic about heaving this telescope around the dome, especially around the pole (that maneuver dreaded by us all). Worse than requiring mere fortitude, the Crossley demands foolhardiness. The telescope is surrounded, on three sides, by a skeletal platform, a couple of planks wide, railed on the outside. In the depths of a moonless winter night one can all too readily stumble off the stairs, walk into the solid iron telescope tube, or plunge towards the base of the mounting. Wariness and second sight are useful faculties in this dome. For all that, the Crossley has been the traditional workhorse of the Lick Observatory, and its PhD machine to boot. It is a curious thing that the existence of the 120-inch has attracted so many young astronomers to Lick, graduate students and postdoctoral astronomers alike, yet only a faculty astronomer can be allotted time on the instrument. In practice, this obstacle is circumvented by alliances between young and established astronomers and the sharing of 120-inch time, both vital and useful practices. But the Crossley smacks of *real* astronomy!

It is cool and dark. Our feet shuffle on the narrow iron stair as we descend below the floor. A switch clicks. Melancholy light washes thick shadows behind the iron pillars that ring this sepulcher. In the center a rectangular brick structure bearing a plate proclaims the last resting place of James Lick, one-time miser, latterly philanthropist (with a little help from his friends). Above this gloomy vault, with its elegantly engineered iron ceiling supports, stands the 36-inch refractor. This is the second largest refractor in the world; Yerkes has a 40-inch. It is mounted on a gray iron rectangular-sectioned column, with a stair spiraling around it. It is unbelievably long but, decades ago, the long-focus refractor was the norm and f/20 was not at all unusual, like The Great Northumberland of Chapter 2. The dome that crowns this quiet, proper, elegantly wood-paneled building is a masterpiece of slim wooden latticework, with thin metal struts laid over this skeleton. The floor is also a work of art. It is polished wood, laid in concentric circles, with darker wooden rings interlaced whose radii relate to planetary orbits in the solar system. A clamor of subterranean motors and gears resonates in the dome. We are moving, gradually, fairly smoothly. Access to the eyepiece is by means of this movable floor, one of the first representatives of a nowadays common genre.

I have never myself worked with the great refractor, but I recall spending a half night with a friend, a man whose perspective of astronomy spans many more years than my own. He is also the only astronomer with whom I am acquainted who is attacking directly the problem of stellar masses. His bread and butter is bifilar (two-wire) micrometer measures of close visual binaries. Once enough accurate measurements are available for each binary it is possible to construct an orbit for the system. If an estimate of distance is possible, for either star, then the angular dimensions of the orbit can be converted into physical sizes. The masses of the stars then become determinate. The work is painstaking; many binaries need to be observed at least once a year, preferably on nights of excellent seeing; and the measurements must be repeated over decades, so slowly do most stars revolve about one another. In recent years efforts have been made to develop faster, more objective (no pun intended) instrumental techniques for visual binary observation. Nevertheless, most of the work is still carried out by a small number of astronomers armed with visual micrometers and with access to long-focus telescopes that provide large image scale.

Lick Observatory stands with one foot firmly rooted in the historic tradition of astronomy, with the great 36-inch refractor and the long-continued astrometric survey of the sky, to determine the stellar proper motions of millions of stars; and the other equally determinedly on the current frontier, with studies of quasars and galaxies at high redshift, all carried out on the 120-inch telescope.

WHITE MOUNTAIN 9

Flying in a helicopter always has a dreamlike quality to me. One is suspended in the air, gliding across the terrain in slow motion apparently through sheer force of thought. Below, the brown hot Owens Valley recedes, with its riverine wrinkles and fields dotted with fruit trees. The barrier of the White Mountains looms ahead as the chopper blades whack the frosty air. Gorge and cliff pass beneath us, virtually unmarred by even a dirt road. Limber and bristlecone pine peep from under thick smooth snow. Ahead, a black Quonset hut stains the landscape. We set down beside the building, unload the rope-net racks above the helicopter's landing skis, and race for cover. The snow whirls past us; like a great bird flailing the cold thin air, the machine climbs and is whisked away. As the clatter of its rotors recedes we are left in silence and remoteness. Coyote tracks cross the oozing whiteness. Peaks in Nevada peer over a hazy horizon while California's mighty Sierra Nevada range flanks us on the other side. It is very still; the quiet roars weightily in our ears; we are compressed to the kernel of self.

We are a short distance from the White Mountain Observatory, established only in 1976 for a very specific type of astronomy. The dome is at almost 13,000 feet, and most of the ocean-born moisture from the west is deposited over the granite body of the Sierra Nevada. Little moisture crosses the Sierras and the arid Owens Valley to fall upon the White Mountain range. The infrared portion of the spectrum is vast, from around one micron (10,000 angstroms) to one millimeter, but most of the science I have alluded to thus far has been

A helicopter prepares to leave for White Mountain Observatory. In the background are the Sierra Nevada mountains.

White Mountain Observatory and the Sno-Cat vehicle used for access.

shortward of twenty-five microns. There is good reason for this restriction. Our atmosphere is quite opaque, for the most part, longward of the ground-based twenty-micron "window." However, there is some transparency at longer wavelengths: around forty microns, a window opened slightly in the past few years from peaks in Arizona and Hawaii; at one hundred microns (from planes and balloons), and around 350 to 450 microns in the submillimeter region.

The site requirements for ground-based submillimeter studies are much more stringent than those for shorter wavelength infrared work. At the altitude of White Mountain, the scale height for water vapor (the physical distance over which the amount of water vapor above us falls by a factor of about 2.8) is some 1,500 feet. There would be appreciable gain, for submillimeter purposes, in the establishment of an observatory on the more than 14,000-foot summit of White Mountain, even by comparison with the present 13,000-foot location. But these rather grubby atmospheric windows are hard to use. There is only a small fraction of winter days and nights when the transparency is good (only fifteen percent at best); the temperature can drop to twenty or thirty degrees below zero Fahrenheit; winds can top 100 miles per hour; and snowfall is substantial. The site is inaccessible except by helicopter for much of the year. Travel between the large, overheated (less so now than in earlier years) Quonset hut and the tiny cramped dome is by tracked snow vehicle. No toilet yet graces the dome. For company there are coyotes, bighorn sheep, lichen-daubed piles of rock, and the splendor of the Sierras across the valley.

The telescope has had a distinguished career. The first major survey of the infrared sky was carried out by Caltech in the 1960's, using a custom-built telescope on Mt. Wilson above Los Angeles. At the core of this instrument is a spun epoxy primary mirror (liquids spinning adopt a paraboloidal surface) overcoated with aluminum; chopping is by rocking this dish from side to side. In 1976 the instrument was relocated on White Mountain and another step was taken in the infant career of submillimeter astronomy. It takes a hardy and strongly motivated person to work here, and it demands great patience. In 1978, in readiness for the coming winter season, an upgrading took place of the Quonset accommodation and of the pointing capability of the telescope. With more comfort for the observers and the equipment improved, there are still the difficulties of high altitude to contend with as well as the unpredictability of good

Above: the lightweight epoxy primary mirror for the 62-inch telescope was originally built by Caltech for an epoch-making sky survey at 2.2 microns in the near infrared.

Left: the submillimeter cryostat at the f/1 prime focus of the 62-inch mirror, which is covered by metal sectors when not in use. Compare with the picture on page 100.

submillimeter nights. It was, in fact, an awful winter for astronomy.

But the scientific potential of the submillimeter region is grand. Much of the galaxy consists of huge, cold, gas clouds far from the localized heating effects of stars. There is dust mingled with the gas at temperatures of around ten degrees absolute. These temperatures promote emission principally at submillimeter wavelengths; thus, a new view of the galaxy would become available, revealing the distribution of the dust component in space. There is always the possibility, when a new spectral window is opened, of discovering a class of objects either previously unknown or with totally unsuspected quantities of emission within the new wavelength window. As this type of astronomy matures, we shall see just what it can pull out of the hat. Until then, people will fly their single-engined planes across the Sierra Nevada, transfer to a helicopter in the Owens Valley, and bring their snowshoes and warm down suits to White Mountain.

10

PALOMAR

Neither words nor pictures can adequately convey the impression of the immensity of Palomar's giant 200-inch reflector. You have to stand underneath the mirror, looking into the Cassegrain "cage," to appreciate what sixteen feet eight inches of optical diameter means (or five meters if you prefer). The cage is just that — a volume of meshed wire surrounding large pieces of equipment that operate at the Cassegrain focus, permitting access to this equipment for the purposes of tweaking electronics or stuffing dry ice in the maws of hungry image-tube coldboxes. It is also roomy enough to accommodate a couple of Volkswagens. This is an enormous telescope, even if the Russians now have an even larger one (a 236-inch reflector).

I stood below the telescope on a warming July morning in 1973, marveling at the immensity of the dome and its gray-green occupant, as Venus walked slowly over the twilight. A famous French spectroscopist needed an assistant in his near-infrared program to study bright stars and planets at extremely high spectral resolution. I was privileged to be his helper for some two nights and three morning twilights. The spectrometer was a marvel — not a world-beater in absolute size, but for a device that traveled, a definite champion. We were installed in a strange level, below the dome floor, but above the usual concrete coudé gallery with its array of expectant mirrors of various sizes. The beam from space entered the giant dome, rattled around three to five mirrors in the telescope, then plunged through a hole in the floor en route to the coudé room. Another mirror had been inserted into the beam, however, and directed it into a large wooden packing crate. Inside this crate lurked one of the most ingenious optical devices that I have ever encountered. This instrument, and its predecessor, has been responsible for the detection of minute quantities (one part per billion!) of certain chemicals (hydrochloric and hydrofluoric acids) in the atmosphere of Venus. It ate a very specific region of the spectrum, from about one to 1.7 microns, and it derived incredibly well-resolved spectra of molecules, breaking the spectra into wonderfully intricate patterns of lines and series of absorption features. If it had a problem, it was that very few people in the world could calculate, in the tremendous detail necessary, the theoretical patterns from molecules radiating under a variety of physical circumstances in order to match theory and observation. It was *that* powerful a machine.

One of our targets was to confirm an earlier weak detection of

Pierre Connes' electronics rack for the near-infrared Michelson interferometer used on the 200-inch Hale reflector. The second oscilloscope from the top shows the spectrum accumulating in real time for one of the four selectable spectral windows.

The Palomar 200-inch observing Venus in the early morning twilight. The Cassegrain
"cage" is well named.

oxygen molecules on Mars, and to seek these on Venus too. Let me describe the process of observation, and you may find that your pre-conceptions about big telescopes are unfounded. Firstly, anyone who works at the coudé focus of a telescope, as opposed to the prime, or Newtonian, or Cassegrain, is buried somewhere in the dark, usually below telescope level, and is isolated from the rest of the dome and from the person driving the telescope. By definition, the coudé is a bent light path, always very long. The beam of radiation that strikes the final subterranean mirror in a coude system is convergent, but only just so! The coudé is designed for extremely long focal length, which, after dispersion in a spectrograph, amounts to exceptionally high spectral resolution. It can be used with conventional photographic plates, or plates behind image tubes for increased sensitivity, or to feed any giant static piece of equipment that does not take kindly to being attached to the back of a telescope and swung around a dome for hours on end. We used the Palomar coudé in this latter role, as a means of feeding a substantial quantity of radiation (hence the need for the 200-inch mirror) into a delicate instrument, carefully aligned and immobile (externally, that is; inside, almost every piece of the equipment slid, swung, trundled, or rotated!). We had a built-in autoguider, that is, a system of sensors that takes some small fraction of the incoming light and tracks on a star or planet once it is acquired and centered. This one small oscilloscope with its jellied image (the seeing, you know) was the only point of contact with the sky, and even this was tens of feet below ground, away from the telescope. An intercom kept us in touch with the telescope operator and was used to signal the end of a spectrum and our desire to move to a subsequent star or planet. There might have been a pair of binoculars at the front end of the light beam, or a 4-inch telescope, or a 400-inch, for all that we could sense from our coop. The bigger the telescope, the more it must function remotely and the more antiseptic does the process of observing become.

Typically, we would observe a single object for a couple of hours at a time. There was remarkably little to do, though many things had to be glanced at from time to time. The tremendous resolution of the spectrometer rendered it impossible to monitor the whole spectral region of interest, so vast would such a display have to be. However, the electronics rack included a special home-brewed device that, in real time, would analyze the totality of data accrued until that moment and would display the resulting spectrum (after performing thousands of calculations, albeit of a simple nature) in full resolution and in four preselected though initially movable tiny spectral windows. One could switch from one to another of these notches in the spectrum to see how matters were progressing. It is in the nature of this type of device, a Michelson interferometer, that the longer one goes on observing, the finer becomes one's resolution. Here we are, taking a spectrum of Mars, with our four windows set to display regions of the Martian spectrum where we would expect to see several different molecular features, including the weak, elusive oxygen emission fingerprints. A green image of Mars is quivering at the center of the autoguider screen; we are ready. For the first few minutes a bright green line snakes and twists grotesquely over the screen monitoring one particular window, then it settles down to much less erratic twitching. We perceive a broad, ill-defined dip in the spectrum. We turn to our novel, after a quick scan of every dial and light in the science-fiction rack of illuminated boxes and knobs. We next raise our eyes twenty minutes into the spectrum to see that the previous feature is now resolved into a series of regularly spaced dips of decreasing depth across the window. The spectrum is still exhibiting St. Vitus' dance, altering even as we inspect it. After forty minutes, these dips are themselves fragmenting into finer features, and the process continues. After an hour and a half the spectrum is still changing, but this is now scarcely perceptible. Between the fine comb of absorption features we can clearly distinguish a series of tiny blips, in emission — the signature of oxygen molecules we had hoped to find.

At daybreak, we would slew the dome to the east to visit Venus. This was permitted, even after sunrise, provided that sunlight did not fall directly onto the primary mirror. The telescope operator (a demon pool player, I hasten to interpolate somewhat irrelevantly) would maneuver the dome and the height of the fabric windshield over the slit to shade the mirror. Finally, I was relieved of my guardianship of the spectrometer and its vast cache of data stored on large magnetic tapes and was free to wander through the dome, or to gaze awestruck at the monster eye. Outside, birds sang, wild flowers shook off the dew and nodded in the sunlight, and below Palomar Mountain the hang-gliding enthusiasts readied their colorful triangles. It was astronomical bedtime.

The "monastery," so styled because of the traditional lack of women and the hermitlike seclusion of the astronomers, is a rambling mansion at Palomar. Meals are served herein following a tradition initiated at the monastery of the 100-inch telescope on Mt. Wilson. The 200-inch observer, nighttime that is, sits at the head of the table with a small bell at his side. It is his responsibility to signal the moment for introduction of the next course from the sweetly odored kitchen into the dining room. The prattle is of quasars, coudé spectra, hypersensitization of near-infrared emulsions. Somewhere on the table, amid butter dishes, meat platters, and juice pitchers is a piece of wood bearing several tiny flags that represent the nationalities of all the astronomers working at the observatory at that instant: a cute touch.

The summits of large telescope domes are good places from which to view the world and philosophize. On my first morning at Palomar I was conducted, by my French senior colleague, along a breathtaking series of internal ladders, curved staircases and steel trapdoors, to a small ledge on the very top of the dome for the 200-inch. The smog of Los Angeles oozed yellow-grey to the north and west; the trees of the Cleveland National Forest formed walls below us; crowds of tourists encircled the base of the dome, unseen, but their voices drifted through the preprandial calm.

While we were allocated morning twilights to observe Venus, another astronomer had the use of the 200-inch during the nighttime hours. Since he was, fortuitously, an acquaintance, there was no problem in receiving permission to look over his shoulder as he worked at the Cassegrain focus. In the early evening we entered the massive prison of the Cassegrain cage to feed, not one or two image tubes, but two banks of sixteen each, part of a low-resolution spectrograph system, loosely akin to the Lick image-tube scanner in sensitivity but considerably poorer in resolution. We minced (literally, in a meat grinder!) chunks of dry ice to feed this malnourished family, and then we retreated (after a suitable dinner) to a small enclosed room on the dome floor, the nerve center for Cassegrain operation. There was a television system here for guiding on galaxies or whatever exotic creatures were under scrutiny, and another TV monitor that displayed the count rates in the thirty-two tubes as each accumulated photons from a (typically thirty-angstrom) chunk of the spectrum. Operation was simple and efficient: point the telescope, acquire the galaxy, guide it into the spectrograph slit, push a button or two, and accumulate information on the spectrum. In fact, so simple was it to observe once an object had been established in the guiding TV, that I found myself, along with an undergraduate student also on his first visit to Palomar, "operating" the mighty 200-inch (while official astronomer and telescope operator pressed their noses to a somewhat livelier television display, in the corner of the control room, where Wolfman Jack performed his Midnight Special). The little 4-inch reflector in England seemed at that moment very very distant, in space, time, and aperture. It was.

The more modern a large telescope, over about 100 inches, the less likely it is to have a dedicated prime-focus cage. Principally, this is a torture device, applied to an observer, for the purpose of direct large field photography. The trend in astronomy is toward remote operation mostly at the Cassegrain focus, and the traditional art of direct photography, while not moribund, is certainly less widely practiced than, say, two or three decades ago. The 200-inch Hale reflector has a prime-focus cage and, like everything about this telescope, it is substantial. We ride the elevator up the inside of the slit in the late afternoon as the dome is cooling off and its hot air is escaping. I feel that it must be far easier to step from the protected platform on the elevator into the cage at night than during the day, because in daylight you can see the gap that must be bridged by foot, leg, and nerve, against the omnipotence of gravity. It is a long way down, past the huge fretwork tube, past the covered primary mirror, to the grey dome floor. For science, we take the step and we are suddenly deeply enclosed. There is room for an observer and boxes of plates here; in fact, there is room for several observers, perhaps a small cocktail party of five or six people (do not mistake factual reporting for hyperbole). Still, it is not a place in which I would choose to suffer an attack of claustrophobia, even if the stars are visible above your head. Or, perhaps, because of the obvious exposure of the cage, would it not be an attack of agoraphobia?

HAT CREEK 11

Like tiny stars winking in the graying dusk, our headlights flashed from the eyes of deer browsing by the road. We were in northern California, not far from volcanic Mt. Lassen, heading for Berkeley's radio observatory at Hat Creek. The two of us were equally inexperienced in long-wavelength techniques, but a colleague, who had been reared as a radio astronomer, had kindly offered to modify the electronics for our desired program and to help set us up on the 85-foot antenna. It was deepest black when we arrived; the elevation was some 5,000 feet, and the stars were brilliant flecks on a velvet cloth.

Next morning our professional colleague tipped the huge telescope to a horizontal position and elevated himself, on a large brightly painted forklift, to the prime focus. It had been difficult to appreciate the size of the equipment at the prime focus until he was working on it. The long support legs converged on a cylinder perhaps six feet long and three feet in diameter. After his adjustments he clambered up a maze of ladders at the back of the dish to complete some final settings which were totally mysterious to us watching fascinatedly from the ground.

All too few hours after this, it seemed, my novitiate friend and I were alone in the quiet computer-lined control room of the 85-foot telescope, mapping the regions of space around a couple of very hot stars in the radio continuum at a wavelength of six centimeters, perhaps the optimal frequency for this dish and its available receivers. The electronic modifications made our assigned task relatively straightforward. In normal operation, a spectrum would have been taken at some tens or hundreds of different but closely spaced radio frequencies. The massive filter bank acted as storage for the signals at different frequencies, which gave information on gas velocities via the Doppler effect. What we wished to do was to operate in a broad continuum bandpass but to scan an extended area in a series of linear cuts across the sky, sampling the signal strength at regular angular intervals. For us, the filter bank stored the spatially separated signals one by one and enabled us to build up our two-dimensional map as a series of scans. Why were we doing this and what were the objects that we were observing?

I had made an interesting identification of a nebula with an infrared source observed from a rocket. Inside this bubble-shaped nebula lay a very hot O star whose optical spectrum indicated that the star was blowing off appreciable quantities of material in the form of a strong stellar wind. The wind seemed to have run into the adjacent gas,

The receiver uncovered at the prime focus of the Hat Creek 85-foot dish.

thereby sweeping out a bubble or cavity in the interstellar medium. The rocket infrared data indicated that there were cool dust grains close to this extremely hot star. We scanned the object from the ground and confirmed the infrared source from Mt. Lemmon. Was this situation of cool dust around such a hot star unique? We located another, seemingly very similar, object from a catalogue of nebulae and observed it at infrared wavelengths. Again there was unexpectedly strong emission. This second nebula was also detected by the Air Force rockets. Now we were intrigued and began to diversify observationally, in order to complete the picture on these two nebulae. We observed the stars and parts of the nebulae at optical wavelengths with the Crossley reflector. There was dust throughout the gas, although our belief was (and still is) that the grains originated from gas in the hot stellar winds. We needed radio maps of both regions to see whether there were other optically invisible exciting stars and to assess the ultraviolet output of each star (since ultraviolet photons

The back structure of the 85-foot antenna.

cause ionization of the gas, enabling radio emission from the liberated electrons). Hence we came to the conclusion that we would have to try our hands at radio astronomy at Hat Creek.

The study was fascinating, and I really had the sense of being some sort of detective. There was a plethora of factual information, drawn from across the electromagnetic spectrum. Could we assemble all these data into a coherent structure, omitting none? It set a precedent in my mind for the great value of the multifaceted approach to any astronomical problem, a precedent to which I hoped to adhere in the future. Operationally, it was not difficult to acquire the data. However, I do confess to a distinctly strange feeling during our run, which began at about four o'clock in the morning, when the sun rose. Outside the window of the control room stood this titan assembly of girders and cables, unperturbed by the daylight. Inside, tapes still spun, calmly storing our data. The twenty-four-hour observing day was a reality!

NRAO
IN ARIZONA

12

It was with some trepidation that I drove up to Kitt Peak again, heading for the National Radio Astronomy Observatory facility that co-exists, just below the summit, with the optical telescopes. This was to be my first solo run as a radio astronomer. On a lower ridge that trails away from the summit, capped by its 158-inch telescope, is a huge dome. What makes it seem so huge is that it resembles a "real" optical dome but on a grand scale. True, it is made of a layer of brilliant white fabric stretched glossily over a metal skeleton, but it has a slit that opens like a conventional dome, rotates, and is inhabited by a telescope. And what a beautiful instrument is housed inside.

The sun shone warmly out of a blue October sky as I parked by the mobile homes that provide dormitory space for visiting radio astronomers. A dull trundling and squeaking sound accompanied the rotation of the dome as I started up the long ramp. Eleven meters (or thirty-six feet) in diameter, like a vast white flower on a trunklike stem, stood the NRAO millimeter-wave antenna. It is an altazimuth mounted dish, consisting of a surface of aluminum plates, ingeniously supported by a web of metal struts. In order to manufacture a telescope for wavelengths of a few millimeters, one must configure a surface to a typical local error of less than 0.14 mm. Further, one must maintain this accuracy as the telescope tilts across the sky, as gravity moves with respect to the dish. Admittedly one can grind optical materials to tolerances of only millionths of an inch but, in the case of a radio telescope, the techniques for engineering are quite different. This is a world of metal struts and plates, of mirrors meters

Above: the 36-foot dish of the National Radio Astronomy Observatory in its giant fabric dome on Kitt Peak.

Left: the prime-focus receiver of the 36-foot telescope. Radiation enters through the square downward-looking orifice; the bulky snake at upper right provides refrigeration.

Inside the control room of the 36-foot dish. The chart recorder at right clearly shows square-wave signals, but these were from quasars, not the author's young stars!

across, of structural deformations induced by gravity and wind. In brief, this is high technology and it is one of the reasons that microwave radio astronomy became an established discipline only within the past few years. This was not by any means the biggest radio antenna I had encountered — I grew up in England only twenty miles from the 250-foot Jodrell Bank Mark I radio dish. However, it was far and away the biggest dish that I had met inside a dome.

For my first couple of hours of orientation, prior to my actual observing period, I simply ogled the beast. I gazed at the spidery prime-focus support legs, at the compact prime-focus receiver and the dark orifice that admitted the radiation. Cable sheaths clung to the support legs, hung in loops and well-organized twists about the telescope. A giant orange snake provided refrigeration at the prime focus to cool the electronics in receivers. Above me, the telescope gyrated to acquire a new object. I left the dome and walked into the control room.

Racks of electronics lined the room; a chart recorder unfurled its inky messages; tapes spun on the computer that takes care separately of telescope tracking and of data handling. Everything was calm and there was an atmosphere of serene control. In conversation with other radio astronomers I realized that there is a world of difference between the level of frenzy that accompanies line observations (for example, emissions due to molecules such as carbon monoxide and formaldehyde) and the relative peace of continuum work. The 36-foot antenna was the principal telescope used for molecular-line astronomy for several years. It is only now beginning to cede this dominance to a handful of smaller, but more accurately figured, new antennas that are more effective at the shortest wavelengths.

If one envisages a molecule as consisting of solid balls attached by springs to one another, then it is clear that such a system can store energy in various and quite different modes. Springs can bend, twist, and stretch. Each such mode is associated with radiation in a specific spectral region, and combinations of modes also can occur. The growth of molecular radio astronomy at millimeter wavelengths took place because of the rotational energy transitions of molecules. What was required was a combination of a dry site (to avoid atmospheric water absorption so troublesome at millimeter wavelengths), accurate engineering (to insure smooth mirror surfaces), and sensitive low-noise electronics and detectors (to sense the tiny interstellar signals above the internal hiss of receivers and amplifiers). In a remarkably short space of time, a highly competitive science arose, with laboratory physicists predicting the precise frequencies of molecular transitions, and radio astronomers bearing these predictions to the telescope for testing in those interstellar laboratories, the dark clouds of gas and dust. Initially, diatomic molecules were detected (e.g. carbon monoxide), then triatomic (e.g. hydrogen cyanide) and, by 1977, polyatomic species with as many as eleven atoms had been seen. The principal difference between line and continuum radio work is that the frequencies of molecular transitions are known so precisely that one needs to correct, almost constantly, for the varying radial velocity of a source relative to the earth as the planet spins. Continuum measurements operate over such broad spectral regions that this problem never arises. I had come to work in the continuum, although at high frequencies (wavelengths of nine and three millimeters). The reason for my donning a radio astronomical hat was a rather curious one.

During the massive optical and infrared survey of young stars carried out at Berkeley, we had discovered that some few percent of the objects showed very strong optical emission lines to such an extent that we could not detect absorption features between these. In short, we could not characterize these stars by spectral types, that is, by temperature. It occurred to me that perhaps these represented young stars significantly more massive than the sun. In fact, using the usual galaxy-averaged mass distribution (how many stars there are of different masses), one could determine what fraction of stars in a random sample would have a mass greater than, say, five solar masses. This proportion was the same as that of the peculiar stars without visible absorption features in our sample. If they were sufficiently hot stars, then one would expect their strong ultraviolet radiation to ionize (split electrons off atoms) the gas in their vicinities. This ionized plasma could then be recognized by its radio emission. My idea was to study the "continuum stars" at two radio frequencies, in order to define the radio spectrum, with the hope of learning whether they were in reality hot high-mass stars. Since the nurseries of young stars that we were studying all lie in the plane of our galaxy, where the low-frequency radio background is high, I opted to work at higher frequencies to reduce this confusion. My feeling was that any high-frequency signals would be sufficiently rare in these regions that they would certainly be due to the stars included in my radio beams. So I left for Tucson.

During my first trip on this program I was scheduled for the nights while another astronomer used the daytime. You know, radio astronomers really do things properly. They brook no interruptions to their round-the-clock operation when the sun rises or sets. Their computers keep the telescopes pointing accurately, tracking at celestial rate, and continuously store the data acquired for subsequent analysis by the observers. There is an operator who points the dish and takes care of the "housekeeping" routines associated with the mechanical and electronic operations. The observer is seated at a separate computer terminal through which he may assemble and add together all his accumulated data. Of course, to some extent, the observer is redundant. His principal decisions are which objects to observe, in what sequence to order these, and how much time to spend looking at each object.

I really wanted to get some good results. I had what I took to be a sensible and feasible study (as did the committee who allocated time to me on the telescope) and, despite my inexperience with radio techniques, I was enthusiastic and conceptually prepared for the work. I detected hardly any of my stars, however! What was even more frustrating to me, and what really highlighted the giant step that is taken when one goes from optical/infrared wavelengths into the radio region, was the work of the daytime observer, with whom I played "Cox and Box." He was studying continuum radiation from quasars over a period of years, seeking sudden changes and hoping to learn whether such changes occurred at all radio frequencies or preferentially at some. It would have taken me a significant length of time to have studied any of his objects in the optical or in the infrared. However, the daytime chart records showed elegant square waves where the man's quasars had stood in the beam and smiled at the radio receiver. On the other hand, my stars, so bright and readily detectable at short wavelengths, could not be seen at radio frequencies. My chart records were harsh, jagged ink smears that marched angrily along the graph paper with absolutely no systematic patterns indicative of even weak radio signals. That, more than anything else, gave me a feeling of groping blindly, in a way that even my infrared studies did not, although they too involved the electronic detection of otherwise invisible radiation. Nevertheless, I enjoyed the radio experience, learned from it, and wished to return when a more sensitive receiver became available.

Several months later, I arrived to find Kitt Peak in a dense cloud cap. Visibility while driving up the mountain road was as low as five yards near the top. Conditions were so poor that I did not even realize that I had driven beyond the short access road to the radio dish until, through the murk, I glimpsed a giant gray doughnut. This apparition represented a foggy view of the life-size concrete model of the giant 158-inch mirror in the biggest Kitt Peak telescope. I turned the car around and crawled the mile back down the road to NRAO. On this trip I had decided to concentrate my efforts on a nine-millimeter study of my stars, using a new receiver that promised a significant gain in sensitivity at that wavelength over the previous ones. The remarkable penetrating power of this almost-centimetric wave was brought home to me vividly on my first night. Actually, this time I had been allocated 105 hours continuously on the dish and had brought an assistant with me so that at least one of us would be present around the clock. On the first night I was very busy and actually detected a star or two (really not that surprising as these were peculiar objects even by my own highly peculiar standards). I phoned up to the diner and requested a couple of sandwiches for midnight lunch. I fled at almost 1 a.m., crept up the road without headlights (for fear of disturbing the optical people on the KPNO telescopes), and raced into the diner. The room was full of despondent faces of optical and infrared astronomers alike. I had been working for several hours through a fairly substantial cirrus overcast without having realized it. The ice crystals aloft were little or no trouble at my relatively long wavelength. Oh yes, I can recommend working at nine millimeters!

As I think back on observing at the 36-foot, several events, procedures, and peculiarities come to mind. Driving down to the dome after midnight lunch, one's first view of the telescope is of a dull red glow that suffuses the dome and pours out of the slit. There are reasons for this which are not mere aesthetics. There is a need for general illumination in order to monitor the position of the dish with respect to other pieces of machinery in the dome. Sometimes a lamp is located up at the prime focus and is directed at a transparent entrance lens (a real, optically designed, see-through lens that performs on radio waves) in order to prevent condensation of moisture and the resulting microwave absorption. Clearly, the other users of Kitt Peak would not take kindly to a bright white light, hence the subdued red illumination. I do not intend this in a disparaging sense, but continuum radio astronomy on faint sources is apt to be boring (or, alternatively, nerve-wracking) if you do not bring with you a mass of reading material or old data requiring analysis. So automated and well-conceived is the operation that even a novice like myself can do radio astronomy at NRAO, at least on this telescope.

If you would care for a number to ponder upon that may illustrate why one must sit, often for hours, in order to detect a single source, then consider the following. Some of the weakest sources that I detected corresponded to a radio signal equivalent to raising the antenna's temperature by only 0.0006 degree Kelvin! Of course, antenna temperature is an observationally oriented and instrumentally specific parameter. In practice, one converts this into a flux of energy at the receiver. Calibrations for this conversion can be established by observing the planets, whose signal strengths can be computed theoretically, and can be checked against one another. Similarly, one uses some bright radio sources as standard candles to check the stability of performance of the system during a run.

I have disguised a very serious consideration in the above sentences. How does one point a radio telescope? Well, the approximate pointing of the dish through its overlord computer is known. One points at a planet, for example, and requests the computer to sample the signals from five positions, the initial supposed planetary location and four others removed by one telescope beamwidth from the initial location in the four cardinal directions. Hopefully, the biggest signal is detected from the initial central position. If not, then the computer suggests where the peak signal should be in order to reproduce the four observations around the central one. The little map is repeated until the central location yields the maximum, and expected, planetary signal. This establishes our pointing to well within a beamwidth, that is, to the order of better than an arc minute. These pointing corrections vary somewhat according to the tilt of the dish, that is to say across the sky, and ideally are used only as a local guide for some particular orientation of the telescope. I remember needing to do this once for a source very close to the western horizon, using Venus, which was almost setting, as my pointing source. The planet was so low down that it was optically virtually extinguished by the long path length through the terrestrial atmosphere.

In retrospect, the oddest telescopic event occurred early one evening as I was preparing to observe some young stars in the constellation of Taurus, which was just rising in the east. My logical source, to establish the local pointing for this extreme tilt of the dish, was Jupiter, a little north of east and also low in the sky. I was part way through a five-point map when I happened to glance through the window of the control room into the dome. The telescope had apparently run amok. It rejected Jupiter, seemingly intent on some observing program more to its own taste. The dish was swinging over my head, through the north towards the west, and at high speed. Alarmed, I called out to the telescope operator. Imperturbably he smiled at me and explained the situation. Inside the trunk of the telescope a mass of cables was coiled around the azimuthal axis. Clearly, one could not go around the sky, rotating this way and that, without making sure that this cable sheaf was not being systematically wound around the axis in one direction. Neglect of this consideration could lead (and once did) to a catastrophe when the bundle of cables snapped or stretched internally. Consequently, to the all-seeing master computer was delegated the task of keeping track of azimuth and never permitting more than one revolution of cable to be accumulated. This task it performed simply by never allowing the telescope to track an object through the direction of due east. Instead, it would rotate the antenna through 360 degrees and pick up the same object slightly south of east if it originally was pointing slightly north but needed to track through east. In my case, Jupiter had risen through due east and the data collection had been terminated. Fascinating creatures, computers . . . .

On a visit to Kitt Peak in late 1978, wearing my infrared astronomer's hat, I encountered cloudy conditions, a regrettably all too common experience in Arizona these days. Several of us decided to drop in on the 36-foot and surprise the observers by our sociability. They, of course, were able to work efficiently through the smooth overcast that stopped optical and infrared operations. I had told my colleagues who accompanied me on this visitation all about the "nutating subreflector," the radio version of the infrared chopping secondary mirror, and about the madhouse that often accompanies line studies, as opposed to continuum. We drove to the dome, beaming pinkly at us in greeting. Poking up was a white conical apparition. Underneath was the 36-foot antenna, but it was covered from prime focus to dish with the world's largest plastic bag, replete with drawcord to tie this bonnet under the dish's chin! So much for seeing the secondary mirror, and, for that matter, the "Cassegrain" receiver, which is actually mounted on top of the primary surface, rather than below. The reason for this enormous party hat was to enable the observation of objects close in the sky to the sun without the resulting, undesirable, direct heating of the metal primary surface. If heated, the dish requires a couple of hours of cooling to resume its normal figure, during which time it is apt to be out of focus because not all the secondary support legs expand to the same extent, thereby tilting the radio axis of the system, and differentially distorting the surface. When we entered the control room I was expecting to find a hive of activity. All was tranquil. The telescope operator had his feet up and a phone to his ear. Perhaps even molecular-line radio astronomy had become the smoothest of routine operations.

The funniest (well, tragi-comic?) tale I heard associated with the NRAO site in Arizona related to the death of a skunk. Apparently, a large group of visitors was resident in one of the two dormitories when a dying skunk chose to shuffle off its mortal coil precisely underneath the dormitory trailer. To say that the odor was lingering and all-pervading was to report matters very quietly. The entire trailer had to be fumigated, and its astronomical occupants bore a quite distinctive malodorous perfume for some considerable time thereafter — all in a night's observing.

13 HAWAII

Craters poke skyward in crumbles. Their shadows rest darkly upon the orange and pink ground. Wearily we force one foot after another until we are standing at the summit. Bright spots pepper our vision; lungs wheeze deeply; pulses race crazily. It is easy to fantasize that one is on the surface of the Moon, but in reality the location is atop Mauna Kea, a volcanic peak on the island of Hawaii almost 14,000 feet above the Pacific. This is close to the ultimately desirable infrared site: high, extremely dry, cloudless, quite accessible, and festooned with telescopes and domes.

What did we go to Hawaii to achieve? Cool stars, on their way to oblivion, puff out thick clouds of gas that condense into dust grains with temperatures of several hundreds of degrees. Dust at these temperatures radiates most of its energy at ten and twenty microns, and a colleague of mine at Berkeley built a complex infrared spectrometer to study this radiation. Our aim was to examine some of the brighter dust shells at high spectral resolution, seeking molecular fingerprints that could enable quite specific identifications to be made of the species of grains that surrounded these dying stars. The mode of observation in this type of work is to observe the stars and to compare their spectra with spectra of the moon taken through an equally long line of sight through the earth's atmosphere. The twenty micron region of our atmosphere is considerably cut up by water vapor absorptions. Since the shape of the lunar spectrum can be predicted, however, it is possible to recognize these terrestrial absorption features and remove them from the spectra of the stars, leaving only

A view from the volcanic summit of Mauna Kea overlooks cinder cones. Far below is a layer of clouds.

true stellar characteristics. From the high altitude of Mauna Kea one can both minimize the depth of these absorptions and insure relatively stable sky conditions to justify the comparison of lunar and stellar spectra taken at somewhat different times.

We flew into Hilo, the airport for Big Island, rented a four-wheel-drive truck, and set off for Mauna Kea. Steadily the rural jumble of Hilo's suburbs fell below and we were driving through a seemingly tropical rain forest. It even rained on us and, since our priceless spectrometer (actually $20,000) was riding in luxury on the passenger's side of the cab, two of us rode bouncingly and windily in the truck bed, swathed in raincoats. Eventually we broke through the cloudy shield that covered Hilo and the lower elevations of the island. The sky was lucid and blue as we bounced across the great broken lava fields of the Kamehameha flow of over a century ago. At the landmark of the Humuulu sheep station we left the road and rose dustily on what I suspect is now a superhighway, but in 1973 was a dirt road. We donned dust masks that covered nose and mouth. The truck turned an orange-gray hue. At every pothole all eyes were on the spectrometer box. Would the delicate (despite its incredible weight) instrument survive the journey from San Francisco? We had

deliberately arrived several days ahead of our scheduled observing
period in order to set up and test our complicated equipment in a
laboratory prior to installation on the 88-inch telescope. In fact, it did
travel well and, once we had acclimatized to the singularly rarefied air
at the summit, we ran through our tests efficiently and successfully.
Of course, this spirit of scientific endeavor totally eluded a certain
airline (the largest in the free world, they claim), which for three days
left our entire electronics racks on their wooden shipping palette
locked up somewhere in the bowels of San Francisco airport. (Upon
our return to California there ensued a pungent correspondence.)
Eventually, by dint of our boss standing over the airline personnel at
San Francisco as the equipment was loaded onto a plane, we were
reunited with all that we needed to set up our data collection system.

Each day was the same. We arose around noon in the echoing
prefabricated building in which all visiting astronomers were accom-
modated. It vibrated constantly to the tune of individual air
conditioning units in each room. Outside it was very hot, very dry,
and very bright. We ate well at the observatory's diner that provided
dinner and a packed lunch for midnight munching. Hawaiian pigs
fled beneath our feet as we crunched over the dry volcanic debris (I
think we ate most of these porcine pups). Slowly, day by day, we
increased the length of time that we spent on the summit where all
the telescopes are located. A good deal is talked about the effects of
high altitude. There are really no generalities to utter. It is almost
totally an individual issue; some people have headaches, others feel
lethargic, still others never seem to notice the lack of air except for
occasional shortness of breath. I manifested a little of each of these
syndromes but mostly felt fine.

At last it was our turn. During the day we tested everything on the
back of the telescope; our spectrometer still worked well. After dinner
we drove up in convoy — in case one vehicle breaks down — with the
night assistants, men whose job it is to protect the telescopes from the
fiendish astronomers whose only concern is their data and not the
longevity of observatory instruments. As the dust settled behind us,
the caravan sneaked through the series of connected potholes that
represented the road, for want of a better term (maybe even this high
altitude section is paved now). The sun was setting as we stood
outside the dome for the 88-inch, facing east. The countersunset
glowed with subdued tints, and the dark triangular shadow of Mauna
Kea lay over the island and its sea of cloud and climbed into the

Looking east from Mauna Kea as the sun sinks, one can see the triangular shadow of the volcano rise into the shadow of the earth.

gray-green earth shadow. A full Moon rode quietly above the mountain shadow, awaiting our observations for calibration of the terrestrial atmosphere that night. In the southwest, Halemaumau grumbled and flickered to itself in volcanic insomnia. The sky was very dark and very clear.

The massive solid tube of the 88-inch telescope (unusual for such a large instrument) stood peacefully in its yellow coat of paint. The first night began. I was in the dome, riding the platform that carried vacuum pumps and racks of electronics up and down and rotated to gain access to the spectrometer at the Cassegrain focus. Down in the oxygenated control room my colleague who had built our machinery watched the data and transferred them to magnetic tape. We had come a long way and had invested a not inconsiderable number of dollars in the trip. In case of spectrometer failure we had even brought with us another piece of instrumentation, an infrared photographic image tube, and a third person to run it if we needed a back-up program. So worried were we about exactly what had been written onto our magnetic tapes, and how accurately, that we had run a test before the first night on the telescope and had flown the resultant tape to California for immediate analysis on the big computer at Berkeley. We repeated this rapid tape freighting with

The solid-tube 88-inch telescope in Hawaii, with the author at the Cassegrain focus where an infrared spectrometer is mounted.

the first night's data. All looked well, according to the feeble crackles that had borne the words of our boss from Berkeley up to the mobile radio-telephone on Mauna Kea. We were where we had planned and hoped to be, making the first observations of this type on our chosen objects at the fabled Hawaiian observatory.

Somewhere close to midnight, as I drove the platform to the ground so that the telescope could be slewed above me to the next object, I would eat my lunch. Unmistakably Hawaiian, unrepeatably delicious: whole wheat bread spread generously with tuna salad that was laden with chopped macadamia nuts and black olives, and sluiced away with a can of "Hawaii's own," guava juice. Then, a rapid trip to the lavatory (or the catwalk outside the dome if it was not too windy), taking care not to slip on the patches of ice that glistened on the dome floor, and it was back to the platform. Not only does the large dome have enormous extractor fans to chase away turbulent hot air that would jeopardize the quality of the seeing, but it also has an internally refrigerated observing floor to reduce thermal gradients further.

The nights were long and wearying. After twelve hours, we would pile into a truck and, blissfully, be chauffeured back to the living quarters 5,000 feet below by the apparently tireless night assistants. I recall the dawns, bouncing nauseatedly across the dusty mountain, trying to retain my midnight lunch that grudgingly suffered my company. On the eastern horizon a faint colorfully flickering light waxed: Jupiter rising. Back we walked to the dormitory, still shaking and clattering to the threnody of air conditioners. And so it was for five gorgeous nights of infrared data. We found evidence for silicon carbide grains around some cool carbon stars, precisely at the frequencies predicted by a Berkeley colleague. We found silicate dust around Betelgeuse and some of its cousins; that was no surpirse, and our dream had been that at our high spectral resolution we would be able to speak of a more precise identification. The universe, however, rarely provides simple honest clues to its inner workings. Suppose there were one type of pure silicate material around a star: that would produce a firm diagnostic infrared spectrum. But what could blur the issue so as to yield the smooth featureless spectra we had detected? Well, the materials need not be pure, and with each chemical impurity the fingerprints are smeared — there could be a range of different types of grain; a range of temperature; a variety of shape, size, and orientation of the grains. In short, almost every realistic, as opposed to ideal, situation would conspire to eliminate the hoped-for sharp, predictable, molecular fingerprints. But it was a successful trip.

Mauna Kea has changed its character since my last visit. There is now scarcely a pseudosummit that lacks a major dome. The site has become internationally recognized, and several endeavors are now under construction. The Canadians and the French, together with the University of Hawaii, are soon to open up a joint 142-inch telescope; the British decided to site their giant 150-inch infrared telescope there; NASA is to erect a 120-inch infrared reflector. And across the valley the volcanoes rumble, hopefully not in earnest.

My most curious memory of Mauna Kea is not at all astronomical. Imagine one of the old silent funnies in which Buster Keaton would flee jerkily across the flickering screen, pursued by clouds of Keystone cops in battered vehicles. You must recall at least one such film in which someone's engine dropped out of their automobile? We were up at 12,000 feet on the mountain, being driven to the telescope on our fourth night of observations, in a staunch but much battered

Toyota Land Cruiser. As the driver revved up and slid into third gear in four-wheel drive, there was a stomach-churning clatter from somewhere below our feet, and the engine died. I looked out through the rear window. As the orange veil of dust settled I saw a long snaky path of black wetness, liberally sprinkled with nuts, bolts, and other pieces of vehicular intestine. Like something out of an old movie, I tell you. I appreciated the convoy principle as six of us piled into a Chevy Blazer to complete the trip to the telescope that evening.

CHILE 14

Closer and closer the circling brown dot comes. It soars across the blue void, seemingly oblivious to the cluster of domes, to the fleet of white Volkswagens chugging along the paved road, to the distant, hazy, snow-frosted peaks. It is a condor: almost one hundred inches of feathered span — an immaculate, proud, soaring machine. Around us, the dead brown glare of desert ricochets off the mind. The desolation and the remoteness benumb the senses. Across our view the ranges of mountains stretch bleak, bright, unwhiskered by vegetation. To the west the ocean is drowsy in the gray-blue haze. Twenty feet away, the condor lands. It waddles cumbrously toward us along a low stone parapet at the cliff's edge. I introduce myself to the giant bird; I tell it of my hopes for the weather, of the science I want to pursue. I speak to it in Spanish, of course, for this is Cerro Tololo Inter-American Observatory, and we are in Chile.

CTIO was established in 1963 as a sister to Kitt Peak. Each is located at a latitude of about thirty degrees, one north and one south. Cerro Tololo is in northern Chile, in the foothills of the Andes and within sight of these giants. It is always an adventure to observe in Chile. The travel is interesting, the culture so different from one's norm, the political and sociological milieu so complex and controversial. Communication is also a challenge. I had never studied Spanish prior to my first Chilean experience in 1975. I vowed that I would do better on my next trip. I did, and in 1977 it was really satisfying to be able to talk with the telescope operators about their life on the mountain, to communicate to them the significance of the piece of astronomy I was

The barren Chilean landscape of Cerro Tololo; in the far distance rise the Andes.

engaged in, and to discuss politics and the realities of life in Chile.

Application for time at CTIO is like the procedure at Kitt Peak — six monthly scheduling periods and deadline dates for proposals three months ahead of the beginning of those periods. We fly into Santiago, the capital of Chile, from the U. S. mainland and overnight in one of the two Sheratons. There is the San Cristobal, on the outskirts of the city; a hotel expensive even by U. S. standards and surely astronomically priced for Chileans, even business executives. The San Cristobal is vast, modern, somewhat hideous, and the temporary home of the local jet-setters. In the heart of Santiago lies the Carrera, a tall, somber old hotel in the European style, slightly faded in its opulence. In either place there is the challenge of the telephone — one really ought to attempt to order from room service in Spanish!

Early the next morning, as the sweet rolls slosh pleasingly in our stomachs amid great bowls of coffee, we are collected by an observatory driver for the 300-mile "shuttle" north to the coastal town of La Serena, the headquarters of CTIO. In 1975 I flew from Santiago to La Serena, but by 1977 this air route had been discontinued. I am glad, for the road trip is fascinating. We leave the sprawling hinterland of Santiago, pass through pretty villages couched in fertile valleys, and then the scenery changes quite abruptly. This coast of Chile is a desert, oddly confronting the Pacific ocean. The road scythes across arid plains, twists past jagged peaks, skirts tiny farms. Above us, the condors wheel and glide. The day is hot as we pause for watermelon, and later for lunch at a restaurant seemingly visited by every bus that plies its trade along this coastal highway. La Serena is inviting in the late afternoon, as we pull into the grounds of the Technical University of Chile and stop outside the modern building that houses the astronomical offshoot. We will spend whatever time we have allocated to last minute preparations for the observing trip, in the astronomers' compound in one of the modern highly comfortable motel-type rooms recently constructed for transients. The reunion with old acquaintances is invigorating, and we are exhorted to present some short informal talk on our work. It is then time for astronomy.

For years we have been spoiled, we of the Northern Hemisphere, by the ready availability of the National Geographic Society-Palomar Observatory *Sky Survey* photographs. For program objects it has always been easy to prepare good finding charts that show the faintest field stars and galaxies. In the southern sky, however, the lack of a deep Schmidt-telescope survey has posed a severe problem. There have always been photographs of especially interesting regions, such as the Magellanic Clouds, taken with South African, Australian, or Chilean reflectors. However, the lack of a total sky survey has been remedied only recently, with the completion of the United Kingdom's Schmidt blue atlas, carried out from an Australian site. The final deep survey is now well advanced, and will be matched, hopefully as rapidly, by a deep red atlas taken with a Chilean-based Schmidt belonging to a European consortium ESO (European Southern Observatory). The blue atlas provides sky coverage adequate for most purposes and it may well be that we will wish to spend a day or two at La Serena, improving our finding charts. Time on the mountain is precious; minutes at the telescope should not be wasted on poor or ambiguous charts, nor on photographs with inadequate image scale.

We decide upon dinner in the big local hotel before forsaking La Serena. The most vivid memory I have of a 1977 meal there is of the coffee that concluded the repast. After all, South America is the home of coffee; if you can't get good coffee there, where else could you? With relish I awaited the aroma of a good strong brew to titillate my nose. A lavishly costumed waiter bore a large salver, with an ornate gold-plated pot in its center, towards our table. With tremendous dignity, he placed thimble-sized coffee cups in front of us and removed the golden hat from the pot. The end of a teaspoon poked curiously over the brim. I craned my neck and verified that the occupant of this aurous vessel was a humble can of Nescafé! South America abounds in anomalies.

The morning is wide and clear as we leave La Serena. The shuttle barrels across a broad and surprisingly green valley whose fertility seems exceeded only by its incongruity amid an otherwise arid landscape. Our route forks from the main road and becomes a well-graded dirt road. We gaze at the contorted scene, in an effort to avoid looking at the speedometer, whose red needle quivers far to the right. The flat plain around us bears the ravages of some titanic calamity — a monstrous flood, perhaps along the empty meandering river bed. Rain here, and the concomitant flash floods seem as out of place in this desert as they do in Arizona, but they are entirely real and just as potent as their northern cousins.

About an hour after we depart from La Serena we are carrying our bewildered suitcase into yet another temporary home, the astronomers' dormitory on the mountain. We are thrown in the deep end. It is lunchtime, and we gather our Spanish, our wits, and our imagination about us as we uncertainly carry a tray to the counter in the diner. It should be said that the food is good: tasty, varied, and quite a blend of Chilean and North American cuisines. I recall, on my first trip to CTIO, being concerned about stomach problems. (This is of distinct importance if you are expected to sit in the prime-focus cage of the 4-meter reflector for most of a night!) In fact, I traveled through Peru and Bolivia before I reached CTIO, and my stomach succumbed to the inevitable malady only the night before I flew to Chile. However, two days of Cerro Tololo's diner and I was restored to normal efficiency. Remarkable, the Latin cuisine.

We are a night ahead of the observing schedule, deliberately, in order to acclimatize to the peak's modest altitude of 7,250 feet and to be present early on the morrow, when the 60-inch telescope is changed from the conventional Cassegrain focus to the new "infrared top-end." As official observers we shall receive the keys to one of the fleet of white Volkswagen "Bugs" that astronomers use to ferry themselves and their night lunches, finding charts, down jackets, novels, and cameras between the dormitory and the domes. The observatory proper lies several hundred feet above the dormitory, accessed by a good paved road. Tonight we walk to the top as our

Cerro Tololo Inter-American Observatory. Some of the reflectors on the crowded summit are, from left to right, the 24-inch Schmidt, 158-inch, 60-inch, and 36-inch.

VW is still the property of the present 60-inch astronomer. Slowly we follow the road, stumbling occasionally in the dark, as the still somewhat unfamiliar beauty of the southern sky assails us.

On the small flat top of the mountain the dark shapes of the domes loom over us and over one another. There was scarcely enough room before the giant dome arrived to house the largest and newest instrument, the sister to Kitt Peak's 4-meter reflector (but with a better primary surface). Now the 4-meter overlooks the neighboring 60-inch, whose view of the southern polar regions is restricted at times by the monster. The Milky Way is a very bright bridge over our heads, with the Magellanic Clouds floating alone like detached puffs of smoke. The colorful splendor of Scorpius, Centaurus, Norma, and Vela arches over the black sky, so dark that the eye cannot grasp its blackness and the brain-eye optics pepper it with internal noise and speckles. We orient ourselves and think about the sun crossing the sky to the north instead of to the south. It's obvious, rationally, but it still feels strange when it happens. The southern sky is incomparably rich; busy *OB* associations clump along the galactic plane; the galactic center region glows like an almost resolved stipple — of course the actual center is optically invisible, buried by untold amounts of obscuring interstellar matter, perhaps one hundred

South polar star trails. The bright star at right is Canopus, and left of center is the Large Magellanic Cloud, visible as a pale elongated patch.

magnitudes of visual extinction, maybe more. We sleep well and arise about noon, making an attempt to move ourselves to a nighttime schedule.

By the time we reach the dome of the 60-inch, the telescope crew has already been at work for quite some time. The new infrared system has just been tested by the staff member who designed and implemented it; it has been made available to visitors and we shall be the first. It boasts an elegant f/30 secondary, only a few inches in diameter, that lives in a fretwork telescope tube of its own. The crew lowers the entire top end of the telescope down through the trapdoors in the dome floor to the large storage area below, the route taken by the primary mirror too on its trips to the aluminizing chamber. Slowly they haul up the infrared tube that attaches to the heavy solid portion of the telescope, in which girdle are mounted the bearings for the declination axis of the telescope. The infrared secondary is tiny by comparison with the massive, conventional, Cassegrain, secondary structure, about two feet across including its sky baffle. But this tiny mirror is an essential ingredient in the type of thinking that underlies "low background" infrared design, which aims to reduce to the abso- lute minimum all the sources of thermal emission contributed by the

The CTIO 60-inch being modified for infrared use. At left a crane lowers the conventional Cassegrain secondary and skeletal telescope tube through the dome floor. At the right stands the infrared top end and tube. Compare the massive conventional secondary housing with the lightweight, superthin, infrared, spider vanes and tiny chopping secondary mirror.

telescope and its optics. It has been an excellent experiment. In 1977 I and a colleague found this telescope, in combination with a fine detector, to be the most sensitive infrared system that we had ever used.

Our study was twofold. The first component was an extension to the southern hemisphere of our big survey of the nuclei of planetary nebulae (between two and twenty microns wavelength) conducted from Mt. Lemmon. CTIO offers us access to the most luminous planetary nebulae and also provides a set of infrared filters whose wavelengths are carefully selected to isolate potential atomic emission lines that arise in the nebulae themselves. Our principal concern was with the infrared continuum emission, as opposed to emission lines. In particular we had found correlations between the radio and infrared properties of the northerly nebulae that can be understood if the infrared continuum is thermal emission from cool (100 to 300 degrees Kelvin) dust grains. Could we substantiate this hypothesis using the bright southern nebulae? Could we strengthen our feeling that perhaps grains are forming in the speedy stellar winds that emanate from the extremely hot (some as hot as 100,000 degrees Kelvin) stars inside planetaries? Could we identify the species of dust grain?

The second aspect of our work also dealt with hot stars losing matter, but this time with the Of stars, the "f" denoting the presence of specific emission lines in optical spectra. These are sometimes almost as hot as the cooler nuclei of planetary nebulae, but are, by contrast, high rather than low-mass stars. Preliminary northern data indicated to us that such stars do not form dust grains. The infrared emission seen in excess over that expected from these stellar photospheres is, therefore, due to gas and as such can be used to determine the quantity of such gas. My colleague was a theoretician who aimed to model the velocity and density fields of the hot gas streaming from the Of stars. Our infrared observations, especially those at ten microns, yield quantitative estimates of the rates at which Of stars lose matter, an aspect crucial to our understanding of their future evolution and likely progenitors. It was an unusual experiment, and not only has it been a successful blend of theory and observation, but a synthesis of ultraviolet, optical, infrared, and radio data! It even worked.

My first trip to Cerro Tololo, in 1975, was more notable for its effect upon my appreciation of nature than for the science achieved. Let me explain. For years I had scurried between San Francisco and Tucson, always managing to be at a telescope out of state whenever earthquakes quivered on the West Coast. I missed several quakes of magnitude four and five, and that only served to make me more curious about the phenomenon and what it felt like to be involved in one. I wonder no more.

It was early morning when I put the 36-inch to bed and topped off the liquid nitrogen jacket around the infrared detector. Exhausted, I drove down the hill in my VW and fell into bed. I had been dozing for a couple of hours when I was awakened by an indefinite sensation. Outside, a truck was passing. It drew closer and closer, and the building vibrated. I got up and peered through the window. At that time some construction was under way to extend the big dining lounge, and the dormitory complex was surrounded by stacks of building materials. In front of me, piles of bricks five feet high danced crazily. There was no truck. It was clearly an earthquake, my first. I settled back to appreciate the phenomenon with the true spirit of investigation. The general noise grew louder. The lamp that dangled from my ceiling swung through an ever-increasing arc. In my bathroom the water slapped and sloshed in the toilet bowl. The shaking grew more violent. At some point in the proceedings I know that fear broke through my hitherto dispassionate observation. What if the quake did not abate? It had seized Cerro Tololo and would not let go. For tens of seconds it continued. I decided against leaving the dormitory in my semiawake state, for fear of being hit by falling masonry or other flying objects. Of course, had I noticed that the astronomers' dormitory was located only about twenty feet back from the lip of an abrupt and high cliff, I might well have left.

My perception of time was grotesquely distorted. The crescendo of sensations was not measurable in seconds or minutes but in heartbeats. After perhaps two or three minutes the violence of the earthquake began to subside. The piles of bricks outside my window moderated their drunken totterings. The water quieted in the bathroom. The lamp in the ceiling continued to swing. I had dressed almost as soon as I realized what was happening. I exited from my room and stood on the ground outside, looking across and along the valley several thousand feet beneath my feet, flanked on its far side by more moderately high peaks. The ground still shook. I had the weirdest sensation of seeing our mountain range moving relative to another, each miles long and thousands of feet high. In the valley, dust hung in the hot air like curtains, no doubt raised by the awesome energy of the quake.

I scurried for my VW and fled up to the summit, praying that the tremors had not shaken my infrared cryostat off the table where I had left it. Behind me the scramble for VWs continued, redolent of the beginning of some ersatz motor rally. In the dome, a glass bottle lay shattered on the concrete floor, but the equipment rested where I had left it only a few short hours earlier. You could not have deduced that there had been a major earthquake except for a weight attached to a steel cord that still gyrated through pendulous arcs from the underbelly of the observing floor above me. For several days and nights there were significant aftershocks, about the size of the typical Californian quakes I had missed in the previous three years. I clearly recall guiding on Gamma Crucis (a bright red, and therefore infrared, calibration standard star) the night immediately following the quake. The star suddenly lurched across the field. One soon became accustomed to such events.

It had been a major earthquake, registering almost seven on the

Richter scale, with the epicenter some distance out in the ocean but around the latitude of La Serena. When I again drove through that town at the end of my observing run, there were still some adobe structures lying in the streets in the form of random jumbles of debris. Injuries were fortunately minimal. Damage was significant but not severe. Chile went about its routine life. On the mountain, the night assistants tracked bright planets in the daytime to ascertain if any telescopes had been moved sufficiently to be pointing no longer at the south celestial pole with their polar axes. No such movements were detected. I know what an earthquake is now, and I am in awe of the energy that is liberated. I hope someone finds a way of taming the San Andreas fault; a quake is not a pretty phenomenon.

The day of the big quake I began to appreciate the remoteness and isolation of Cerro Tololo. As usual, the telephone lines between La Serena and the peak were inoperative; perhaps they had even broken that day. We wanted to raise La Serena for information on the tremors. There is a radio transceiver set up on the mountain too, that can communicate with the La Serena headquarters. It can also cross the equator to call Tucson. We could not speak to La Serena. The radio operator tuned for Kitt Peak. It was a weird combination of events. We were listening to a phone call originating in Canada, aimed at La Serena, via a radio patch in Tucson. Everyone wanted to get news from Chile. At the end of the phone call, we were able to speak to La Serena, via the link to Arizona! A journey of 10,000 miles to bridge a gap of only thirty-five. But that is Chile.

EPILOG 15

With only a modest amount of navigation through life, the existence of an astronomer can be exciting, enjoyable, and almost constantly endowed with the sense of exploration. But exactly what situations should *you* steer for — or equally important, try to avoid — if your mind is set on a career in astronomy? I cannot claim to have acquired an all-encompassing perspective in the past decade, but there are a number of commonsense considerations that I can pass on to you. Let's look at the issue from the viewpoint of somebody in high school.

Astronomy is a specific aspect of physics and as such calls for a familiarity with physical principles. You should be a person who enjoys physics and science in general and be happy dabbling in mathematics. When I teach introductory astronomy I frequently find adults attending my class. Their reasons for being there are a curiosity about and a fascination with astronomy and space exploration; for many this has been a delayed interest. In high school these people almost invariably encountered a poor teacher of physics who inadvertently steered them away from science. Now no one is asking you to become a physicist on the basis of high school courses — you will do it all again in college, in more detail — but it behooves you to develop a feel for the subject. Libraries abound in elementary astronomy texts. Most modern books stress a descriptive, rather than a mathematical, approach and contain the bare minimum of equations. Read these and become familiar with the language of astronomy, and the time will be well invested.

If you are seriously thinking about astronomy, you may well own or have access to a small telescope. Get used to what you can see with it. Over the years you will find yourself learning the sky, almost without realizing that it is happening. Don't specialize — be open to everything that you can observe with your instrument. Watch the planetary moons in the solar system and the subtle markings on Venus and Mars; browse through the open clusters in Auriga and Cassiopeia and the globular clusters and planetary nebulae of the summer months; look for the galaxies in Coma Berenices, Virgo, and Leo. Even if you eventually become a theoretician, you'll have gained an appreciation for the process of observation.

Look around your hometown — is there a government laboratory with astronomical connections; perhaps through your local high school you could get a summer job, or at least an inside look at the research work? Is there an observatory nearby, or a college astronomy department that is advertising for part-time help by high-school students? Such opportunities are extremely valuable for investigating the life of astronomers from the inside, long before you need to decide whether the career is for you.

At college level, don't major in astronomy, even if you take some courses in it. This advice was given to me years ago, before I began university. The idea is not to limit yourself too early to a narrow field. If you concentrate on physics and/or mathematics, you will have a solid foundation for a career in astronomy as well as back-up expertise if you later change your mind. (Some organic chemistry might also be useful, given the rapid development of molecular radio astronomy in recent years.) What is crucial at the undergraduate level is good teaching. You may gain a better, more personalized, education and be happier at a small school, where the instructors are devoted solely to teaching, than at a great campus, with its classes of several hundred and its professors divided between teaching, research, and the continuous scramble for funding.

Choosing a graduate school is a different issue. When you first visit campuses to meet the faculty, spend an equal amount of time talking with the graduate students already enrolled. Ask them if their advisors are generous with their time, or are elusive, or are continually away at meetings. Find out who does theory, who makes observations. See who is proficient at drawing money through contracts and can, therefore, support several students. Look for a department with representation from several different branches of astronomy. Are there X-ray, radio, and infrared researchers as well as optical people? Does the campus have ready access to its own or a nearby observatory, and typically how much time can students obtain at the telescopes, either with a member of the faculty or by themselves? Graduate school is a crucial period in your development — you want to feel immersed in astronomy, to know that the people around you are contributing actively to the growth of knowledge, to be exposed to the multitude of disciplines and techniques that mark modern astrophysics. Graduate school is also the time to be at a large campus, in a big department. If you have tenacity and enthusiasm, you will make it through to a PhD.

The Astronomer Royal of England once told me, "Don't become an observer just because you don't think you're good enough to be a theoretician." I pass this comment on to you. Ideally, you should be both. Your research should involve observations, in whose collection you are intimately involved (how else could you know how reliable they are?), and a theoretical analysis of these. Choose your topic with care and deliberation. Solicit advice from the faculty and from students who are familiar with the equipment you will need to use. Your research advisor will be able to judge whether a developing field will generate thesis-size areas of work. If your campus is one of several pioneering in a new technique, get involved; this is your fastest route to the excitement of astronomy, to the feeling that you are contributing. New techniques also reopen whole fields which have lain dormant, awaiting a fresh approach.

Never become so mired in a piece of work that you lose sight of its relevance from other perspectives. If you feel that to answer a particular astrophysical question needs several techniques, rather than a specialized one, try to get the various observations, or develop the theoretical approaches. Many issues today are amenable to a blend of weapons: radio and ultraviolet — X-ray and radio — infrared and optical and radio — high-resolution spectroscopy and broadband photometry — ground-based mapping with high resolution and comparison with wide-field rocket and/or balloon surveys.

You don't necessarily have to be remembered as the *first* person to make a particular observation or do a type of study. It is better to be labeled as the first to "get the right answer," or to find a fruitful approach, or to lay a solid foundation for future work. The last word carries more weight than the first. Keep an eye on the frontiers, through the journals and discussions within your department and visitors to it, but always look for shaky data and for unsupported speculation. Could something be better measured by a different piece of equipment? Can one theory be differentiated from an apparently viable competitor by some critical observation? Can you obtain quantitative data to substantiate someone else's qualitative idea?

Don't be afraid of voicing ideas that seem offbeat, even a little crazy. Science grows more by inputs from new directions than by mere extrapolation from conventional viewpoints. If you think you can tie together apparently unconnected pieces of astronomy, bounce your view off a few people. Their scepticism and comments will either sharpen and convince you of the merit of your idea, or present you with a contradiction or even a critical test. Some of the most fruitful ideas I've had came over lunchtime pizzas, in flight to an observatory, in an elevator. In this manner you will naturally develop the confidence to draft proposals for observing at places like Kitt Peak, where your program is in direct competition with others for telescope time, and eventually to write full-blooded proposals to national agencies in quest of funds to pursue your research.

Eventually you will emerge with PhD in hand, looking for a job. Rarely does one enter into a faculty post fresh out of graduate school. Most academic positions are advertised for candidates who have a couple of years' experience in postdoctoral research. Now you have some further critical choices to make. It may be possible to stay at the department where you obtained your PhD and to continue in the same general area, either with new equipment or with an instrument that you may have developed for your doctorate. To the world outside your campus you will be regarded as someone with expertise in whatever topic your previous work was in, either through your publications in the journals, or through personal knowledge of you (often a powerful lobby). Your PhD is really only a license to practice research — it indicates a certain potential, stamina, determination, and self-motivation, especially if you obtained it after a modest investment of time (five to six years is typical now). Wherever you get a post-doctoral position, be alert for new directions in astronomy. Find new people with whom to collaborate. Apply your technique to their problems and interests. Broaden your astronomy and select your place of work to enhance your opportunities for this expansion.

Sometimes you will find posts advertised at new observatories, perhaps remote from your country. These can provide exciting positions. You may find yourself obliged to live in a foreign country, or you may find it necessary to live in an isolated environment in your own country — in each case distant from big cities, ready access to culture, or whatever may be to your taste. Nevertheless, for a one to two year period you can survive and accumulate several years of data to analyze in the bargain. Sooner or later we may all be faced with such a compromise of having to move to an undesirable location in order to have a secure job. The alternatives are to join the mobile worldwide population of young astronomers, moving every one to five years from one fellowship position to the next, or to consider leaving astronomy. This is a harsh reality to face, but one that should be given some thought even as you enter graduate school.

If you should want to settle in a given city, state, or country some time in the future, will you have a salable useful talent? You can then fall back on your Bachelor's in physics or mathematics, or on your expertise gained during graduate school with detectors, electronics, cryogenics, rocket hardware, or computing. There are many industries in the real world that might welcome your background in the above areas, especially in computing. Computers are here to stay, and they have only begun to spin their essential threads across society. Get acquainted with them — how to write software, how to get in and out of them, how systems are conceptualized. You won't regret this familiarity in later life. You may decide that you would like to teach full-time, in which case you can try for posts at community or junior colleges, at small private schools, with university extramural boards. If there is a telescope, even a small one, at one of these institutions, your chance of making astronomy interesting to your students will be greatly enhanced, so be sure you know how to use one.

Astronomy is a small piece of science, much at the mercy of external influences, such as the economy, the existence of an active space program, and technological growth. My father has acted as a staunch devil's advocate over the years, always assuring me of the ephemeral and precarious nature of astronomy as a profession. So far I have been lucky; he may be proven right in the future, but I have enjoyed immensely my quest for telescopes.

INDEX